AF587368

TRANSPORTATION ISSUES, POLICIES AND R&D

VEHICLE-TO-VEHICLE TECHNOLOGIES FOR INTELLIGENT TRANSPORTATION SYSTEMS

DEVELOPMENT, CHALLENGES AND SECURITY PROPOSALS

TRANSPORTATION ISSUES, POLICIES AND R&D

Additional books in this series can be found on Nova's website under the Series tab.

Additional E-books in this series can be found on Nova's website under the E-book tab.

TRANSPORTATION ISSUES, POLICIES AND R&D

VEHICLE-TO-VEHICLE TECHNOLOGIES FOR INTELLIGENT TRANSPORTATION SYSTEMS

DEVELOPMENT, CHALLENGES AND SECURITY PROPOSALS

DOUGLAS LACEY
EDITOR

New York

Copyright © 2014 by Nova Science Publishers, Inc.

All rights reserved. No part of this book may be reproduced, stored in a retrieval system or transmitted in any form or by any means: electronic, electrostatic, magnetic, tape, mechanical photocopying, recording or otherwise without the written permission of the Publisher.

For permission to use material from this book please contact us:
Telephone 631-231-7269; Fax 631-231-8175
Web Site: http://www.novapublishers.com

NOTICE TO THE READER

The Publisher has taken reasonable care in the preparation of this book, but makes no expressed or implied warranty of any kind and assumes no responsibility for any errors or omissions. No liability is assumed for incidental or consequential damages in connection with or arising out of information contained in this book. The Publisher shall not be liable for any special, consequential, or exemplary damages resulting, in whole or in part, from the readers' use of, or reliance upon, this material. Any parts of this book based on government reports are so indicated and copyright is claimed for those parts to the extent applicable to compilations of such works.

Independent verification should be sought for any data, advice or recommendations contained in this book. In addition, no responsibility is assumed by the publisher for any injury and/or damage to persons or property arising from any methods, products, instructions, ideas or otherwise contained in this publication.

This publication is designed to provide accurate and authoritative information with regard to the subject matter covered herein. It is sold with the clear understanding that the Publisher is not engaged in rendering legal or any other professional services. If legal or any other expert assistance is required, the services of a competent person should be sought. FROM A DECLARATION OF PARTICIPANTS JOINTLY ADOPTED BY A COMMITTEE OF THE AMERICAN BAR ASSOCIATION AND A COMMITTEE OF PUBLISHERS.

Additional color graphics may be available in the e-book version of this book.

LIBRARY OF CONGRESS CATALOGING-IN-PUBLICATION DATA

ISBN: 978-1-63117-045-4

Published by Nova Science Publishers, Inc. † New York

CONTENTS

PREFACE

The development of vehicle-to-vehicle (V2V) technologies has progressed to the point of real world testing, and if broadly deployed, they are anticipated to offer significant safety benefits. Efforts by the U.S. Department of Transportation (DOT) and the automobile industry have focused on developing in-vehicle components such as hardware to facilitate communications among vehicles, safety software applications to analyze data and identify potential collisions, vehicle features that warn drivers, and a national communication security system to ensure trust in the data transmitted among vehicles. This book focuses on the V2V technologies expected to offer safety benefits as well as the technical description and identification of policy and institutional issues.

Chapter 1 – In 2011, 5.3 million vehicle crashes in the United States resulted in more than 2.2 million injuries and about 32,000 fatalities. While improvements in automobile safety have reduced the number of fatalities in recent decades, DOT has worked with the automobile industry to develop V2V technologies, through which vehicles are capable of warning drivers of imminent collisions by sharing data, including information on speed and location, with nearby vehicles. GAO was asked to review the status of V2V technologies. GAO examined (1) the state of development of V2V technologies and their anticipated benefits; (2) the challenges, if any, that will affect the deployment of these technologies and what actions, if any, DOT is taking to address them; and (3) what is known about the potential costs associated with these technologies.

GAO reviewed documentation on V2V technology-related efforts by DOT and automobile manufacturers, visited a pilot study of V2V technologies in Michigan, and interviewed DOT officials, automobile manufacturers, and 21 experts identified by the National Academies of Sciences. Experts were selected based on their level of knowledge and to represent a variety of subject areas related to V2V technology development.

DOT and the Federal Communications Commission reviewed a draft of this report and provided technical comments which were incorporated as appropriate.

Chapter 2 – This report identifies the security approach associated with a communications data delivery system that supports vehicle-to-vehicle (V2V) and vehicle-to-infrastructure (V2I) communications. The report describes the risks associated with communications security and identifies approaches for addressing those risks. It also identifies and describes the policy and institutional issues that require focus in support of implementation and operations, as well as the balance needed among the priorities of security and safety with cost, privacy, enforcement, and other institutional issues.

The approach described in this report is a first step in identifying the technical and policy requirements that will form the basis for a prototype model that will be tested during the 2012-2013 Safety Pilot Model Deployment, located in Ann Arbor, Michigan. The prototype will be tested along with draft policies and procedures. Results of the test will inform the requirements, specifications, and guidelines for implementing an operational system.

In: Vehicle-to-Vehicle Technologies ...
Editor: Douglas Lacey

ISBN: 978-1-63117-045-4
© 2014 Nova Science Publishers, Inc.

Chapter 1

INTELLIGENT TRANSPORTATION SYSTEMS: VEHICLE-TO-VEHICLE TECHNOLOGIES EXPECTED TO OFFER SAFETY BENEFITS, BUT A VARIETY OF DEPLOYMENT CHALLENGES EXIST*

United States Government Accountability Office

WHY GAO DID THIS STUDY

In 2011, 5.3 million vehicle crashes in the United States resulted in more than 2.2 million injuries and about 32,000 fatalities. While improvements in automobile safety have reduced the number of fatalities in recent decades, DOT has worked with the automobile industry to develop V2V technologies, through which vehicles are capable of warning drivers of imminent collisions by sharing data, including information on speed and location, with nearby vehicles. GAO was asked to review the status of V2V technologies. GAO examined (1) the state of development of V2V technologies and their anticipated benefits; (2) the challenges, if any, that will affect the deployment of these technologies and what actions, if any, DOT is taking to address them;

* This is an edited, reformatted and augmented version of the Highlights of GAO-14-13, a report to congressional requesters, dated November, 2013.

and (3) what is known about the potential costs associated with these technologies.

GAO reviewed documentation on V2V technology-related efforts by DOT and automobile manufacturers, visited a pilot study of V2V technologies in Michigan, and interviewed DOT officials, automobile manufacturers, and 21 experts identified by the National Academies of Sciences. Experts were selected based on their level of knowledge and to represent a variety of subject areas related to V2V technology development.

DOT and the Federal Communications Commission reviewed a draft of this report and provided technical comments which were incorporated as appropriate.

WHAT GAO FOUND

The development of vehicle-to-vehicle (V2V) technologies has progressed to the point of real world testing, and if broadly deployed, they are anticipated to offer significant safety benefits. Efforts by the U.S. Department of Transportation (DOT) and the automobile industry have focused on developing: 1) in-vehicle components such as hardware to facilitate communications among vehicles, 2) safety software applications to analyze data and identify potential collisions, 3) vehicle features that warn drivers, and 4) a national communication security system to ensure trust in the data transmitted among vehicles. According to DOT, if widely deployed, V2V technologies could provide warnings to drivers in as much as 76 percent of potential multi-vehicle collisions involving at least one light vehicle, such as a passenger car. Ultimately, however, the level of benefits realized will depend on the extent of the deployment of these technologies and the effectiveness of V2V warnings in eliciting appropriate driver responses. The continued progress of V2V technology development hinges on a decision that the National Highway Traffic Safety Administration (NHTSA) plans to make in late 2013 on how to proceed regarding these technologies. One option would be to pursue a rulemaking requiring their inclusion in new vehicles.

The deployment of V2V technologies faces a number of challenges, which DOT is working with the automobile industry to address. According to experts, DOT officials, automobile manufacturers, and other stakeholders GAO interviewed, these challenges include: 1) finalizing the technical framework and management framework of a V2V communication security system, which will be unique in its size and structure; 2) ensuring that the possible sharing with other wireless users of the radio-frequency spectrum used by V2V communications will not adversely affect V2V technology's performance; 3) ensuring that drivers respond appropriately to warnings of potential collisions; 4) addressing the uncertainty related to potential liability issues posed by V2V technologies; and 5) addressing any concerns the public may have, including those related to privacy. DOT is collaborating with automobile manufacturers and others to find potential technical and policy solutions to these challenges and plans to continue these efforts. Although V2V technologies are being tested in a real-world pilot that will end in February 2014, DOT officials stated that they cannot fully plan for deployment until NHTSA decides how to proceed later this year.

DOT and the automobile industry are currently analyzing the total costs associated with V2V technologies, which include the costs of both in-vehicle components and a communication security system. All of the automobile manufacturers GAO interviewed said that it is difficult to estimate in-vehicle V2V component costs at this time because too many factors—such as future production volumes and the time frame of deployment—remain unknown. The costs associated with a V2V communication security system also remain unknown as the specifics of the system's technical framework and management structure are not yet finalized. While the costs of in-vehicle V2V components may be modest relative to the price of a new vehicle, some experts noted that the potential costs associated with the operation of a V2V communication security system could be significant. Further, it is currently unclear who—consumers, automobile manufacturers, DOT, state and local governments, or others—would pay the costs associated with a V2V communication security system.

ABBREVIATIONS

CAMP	Crash Avoidance Metrics Partnership
CVT	connected vehicle technologies
DOT	Department of Transportation
DSRC	dedicated short-range communications
FCC	Federal Communications Commission
GHz	gigahertz
ITS	intelligent transportation systems
LiDAR	light detection and ranging
MHz	megahertz

NAS	National Academies of Sciences
NHTSA	National Highway Traffic Safety Administration
NTIA	National Telecommunications and Information Administration
RADAR	radio detection and ranging
RITA	Research and Innovative Technology Administration
VIIC	Vehicle Infrastructure Integration Consortium
VSC	Vehicle Safety Communications
V2V	vehicle-to-vehicle

November 1, 2013

The Honorable Lamar Smith
Chairman
The Honorable Eddie Bernice Johnson
Ranking Member
Committee on Science, Space, and Technology
House of Representatives

The Honorable John J. Duncan
The Honorable Ralph Hall
The Honorable Randy Hultgren
House of Representatives

Every year, motor vehicle crashes in the United States result in numerous injuries and deaths and high economic costs. For example, according to the Department of Transportation's (DOT) National Highway Traffic Safety Administration (NHTSA), 5.3 million police-reported vehicle crashes occurred in the United States in 2011, resulting in about 32,000 fatalities and more than 2.2 million injuries.[1]

NHTSA aims to reduce injuries, deaths, and economic losses resulting from motor vehicle crashes through a number of actions, such as establishing safety standards for motor vehicles[2] and conducting research that supports vehicle safety. NHTSA's safety standards cover various aspects of vehicle safety, such as brakes, headlights, seat belts, air bags, and child restraints. Improvements in automobile safety that aim to protect drivers and occupants in the event of a collision—including features such as seat belts and airbags—have reduced fatalities. Recently, however, the automobile industry has begun to introduce technologies that are intended to prevent accidents. Crash avoidance technologies, which use sensors such as cameras and radar, can observe a vehicle's visible surroundings and issue warnings to the driver when certain types of collisions with other vehicles or obstacles appear to be imminent.[3]

DOT and the automobile industry have been conducting research on new types of technologies to prevent crashes—called vehicle-to-vehicle (V2V) technologies—in recent years. These technologies facilitate the sharing of data, such as vehicle speed and location, among vehicles to warn drivers of potential collisions. Based on the data shared, V2V technologies are capable of warning drivers of imminent collisions, including some that sensor-based crash avoidance technologies would be unable to detect.[4] DOT's efforts related to these technologies are being led by NHTSA and the Intelligent Transportation Systems (ITS) Joint Program Office within DOT's Research and Innovative Technology Administration (RITA). According to NHTSA, if V2V technologies are widely deployed, they have the potential to address 76 percent of multi-vehicle crashes involving at least one light vehicle by providing warnings to drivers. In late 2013, NHTSA is planning to announce further actions it will take regarding these technologies for passenger vehicles, including potentially announcing intent to pursue future regulatory action, such as a proposed rulemaking to mandate the installation of V2V technologies in newly manufactured passenger vehicles.[5] The inclusion of V2V technologies in newly manufactured vehicles could increase vehicle costs.

You asked us to examine the development and possible future deployment of V2V technologies. We examined: (1) the state of development of V2V technologies and their anticipated benefits; (2) the challenges, if any, that will affect the deployment of these technologies, and what actions, if any, DOT is taking to address them; and (3) what is known about the potential costs associated with these technologies for automobile manufacturers and consumers.

To address these issues, we reviewed documentation relevant to the V2V technology research efforts of DOT and the automobile industry, such as DOT's *ITS Strategic Research Plan, 2010 - 2014 Progress Update 2012* and documentation related to DOT's efforts to estimate the potential benefits of V2V technologies. We interviewed NHTSA and RITA officials about these efforts. We also visited the DOT-sponsored V2V Safety Pilot Model Deployment in Ann Arbor, Michigan, and received a demonstration of V2V technologies in the Detroit area from automobile manufacturers working on the development of these technologies.[6] We conducted structured interviews with 21 experts identified by the National Academies of Sciences as knowledgeable in the areas of V2V technology development and interoperability, technology deployment, production of light-duty passenger vehicles, data privacy and security, legal and policy issues, and human factors[7] issues related to V2V technologies. The structured interviews included asking selected experts to rate the extent to which a series of issues—which we identified based on initial interviews with automobile manufacturers, DOT, and others—present challenges to the deployment of V2V technologies. We selected experts who represented both domestic and international automobile manufacturers, suppliers of V2V devices, a telecommunications company, and state governments, as well as automotive industry experts and academic researchers. In addition, we interviewed representatives of 10 automobile manufacturers involved in collaborative V2V technology development efforts, a V2V device supplier, and associations knowledgeable about the development of V2V technologies. We also interviewed officials with the Federal Communications Commission (FCC) about the use of radio-frequency spectrum by V2V communications. Further details about our scope and methodology can be found in appendix I. Our structured guide for interviewing experts is reproduced in appendix II.

We conducted this performance audit from October 2012 to November 2013 in accordance with generally accepted government auditing standards. Those standards require that we plan and perform the audit to obtain sufficient, appropriate evidence to provide a reasonable basis for our findings and conclusions based on our audit objectives. We believe that the evidence obtained provides a reasonable basis for our findings and conclusions based on our audit objectives.

Background

Although there were about 32,000 motor vehicle traffic fatalities in the United States in 2011, the number has generally declined in recent years. According to NHTSA, the rate of 1.10 fatalities per 100 million vehicle miles traveled in 2011 represented an all-time low and was 28 percent lower than the rate of 1.52 in 2001.[8] DOT has attributed reductions in fatality rates to several factors, including increased use of in-vehicle safety features such as safety belts and a reduction in fatalities related to alcohol-impaired driving.

In recent years, automobile manufacturers have begun to equip some newly manufactured vehicles with sensor-based crash avoidance technologies intended to prevent accidents and further reduce the number of fatalities.

These technologies employ sensors such as cameras, radio detection and ranging, and light detection and ranging[9] to observe a vehicle's surroundings.[10] A vehicle equipped with such technologies is capable of detecting potential collisions with other vehicles or obstacles within a range of about 150 meters (about 500 feet) and alerting drivers through applications such as forward collision warnings and lane departure warnings;[11] however, such warnings are limited to the threats within the field of view of the vehicle's sensors.

DOT has worked with the automobile industry and others to lead and fund research on connected vehicle technologies, which include V2V technologies as well as vehicle-to-infrastructure technologies. From fiscal years 2003 through 2012, the funding DOT has made available for efforts on connected vehicle technologies, part of its Intelligent Transportation Systems (ITS)[12] research program, totaled about $445 million and ranged from a low of $17 million in 2008 to a high of $84 million in 2011.[13] In a connected vehicle environment, data is shared wirelessly among vehicles (V2V communications) or between vehicles and infrastructure (vehicle-to- infrastructure communications) using dedicated short-range communications (DSRC), a technology similar to Wi-Fi that offers a link through which vehicles and infrastructure can transmit messages over a range of about 300 to 500 meters (about 1,000 to 1,600 feet). Based on analysis of internal data and data received from other vehicles, a vehicle equipped with V2V technologies is able to issue a warning to its driver when a collision with another similarly equipped vehicle could occur. The range of V2V communications is not only greater than that of existing sensor-based technologies, but due to the sharing of data between vehicles, V2V technologies are capable of alerting drivers to potential collisions that are not visible to existing sensor-based technologies, such as a stopped vehicle blocked from view or a moving vehicle at a blind intersection (see figure 1).

While V2V technologies send and receive data among vehicles, vehicle-to-infrastructure technologies send and receive data between vehicles and infrastructure, such as traffic signals. Vehicle-to-infrastructure technologies could offer additional safety features that V2V technologies cannot, such as providing drivers with additional warnings when traffic signals are about to change, warnings that could help reduce collisions at intersections. In addition, these technologies can offer potential mobility and environmental benefits; for example, they can collect, analyze, and provide drivers with data on upcoming roadway and traffic conditions and suggest alternate routes when roadways are congested.[14]

In addition to federal efforts, automobile manufacturers have formed a number of consortia to research and develop V2V communication-based technologies and have collaborated with DOT on these efforts. Since 2002, DOT has awarded $77 million in funding, under cost-sharing agreements to support projects related to V2V technologies, to a number of consortia established by the Crash Avoidance Metrics Partnership (CAMP). This partnership between Ford Motor Company and General Motors, L.L.C. is currently working with 6 other automobile manufacturers through the Vehicle Safety Communications (VSC) 3 Consortium to collaboratively address safety issues through advancements in vehicle communications.[15] Additionally, the Vehicle Infrastructure Integration Consortium (VIIC) was created in 2005 and is currently comprised of 10 automobile manufacturers with the goal of identifying and promoting policy solutions needed to support the development and deployment of V2V and vehicle-to-infrastructure technologies.[16]

The automobile industry (as well as some technology companies such as Google) is also working to develop autonomous vehicle technologies, which would control steering, acceleration, and braking without a driver's input.[17] Some automobile manufacturers have started to introduce semi- autonomous sensor-based technologies capable of reducing a vehicle's speed through adaptive cruise control[18] or even stopping a vehicle through automatic braking when a collision is imminent. Autonomous vehicles are being designed with the intent that a driver is needed to provide destination or navigation information, but not needed to control the vehicle.

Because V2V communications depend upon DSRC technology to transmit data among vehicles, the deployment of V2V technologies will require use of the radio-frequency spectrum. FCC has allocated spectrum for use by DSRC technologies that are part of DOT's ITS research program.[19] Specifically, in 1999, FCC allocated 75 megahertz (MHz) of spectrum[20]—the 5.850 to 5.925 gigahertz (GHz) band (5.9 GHz band)— for the primary purpose of improving transportation safety and adopted basic technical rules for DSRC operations.[21] In 2003, FCC established licensing and service rules for the 5.9 GHz band to provide a short-range, wireless link for transferring information between vehicles and roadside systems.[22]

However, the President and Congress have responded to growing demand for wireless broadband services by making changes in the law to promote efficient use of spectrum, including the band previously set aside for use by DSRC-based technologies. For example, the Middle Class Tax Relief and Job Creation Act of 2012 required the National Telecommunications and Information Administration (NTIA) to conduct a study evaluating spectrum-sharing technologies and the potential risk to federal users if unlicensed devices[23] were allowed to operate in the 5.9 GHz band.[24]

Given the potential benefits of V2V technologies, NHTSA has said it will decide by the end of 2013 how it will proceed with respect to these technologies for passenger vehicles. According to NHTSA officials, they are considering several possible actions, including announcing NHTSA's intent to pursue future regulatory action, such as a proposed rulemaking to mandate the installation of these technologies in newly manufactured passenger vehicles[25] and continuing to research and develop these technologies for passenger vehicles.

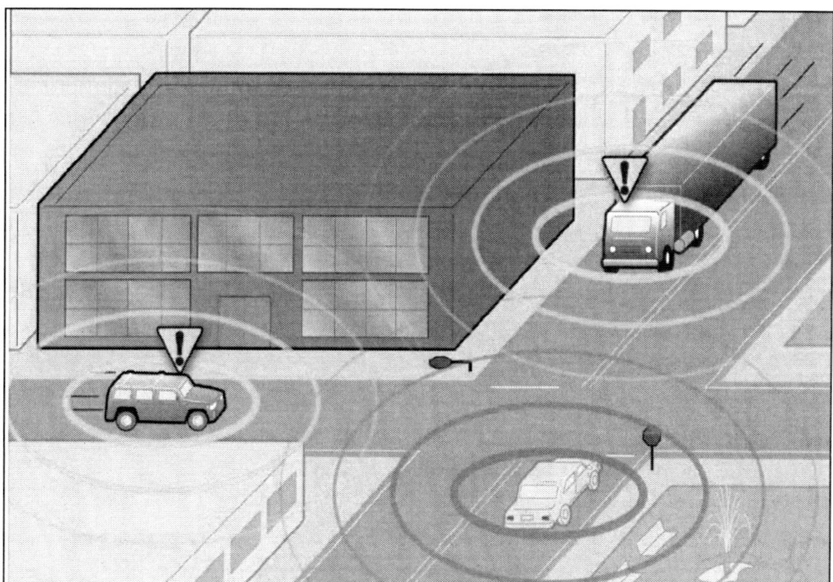

Source: GAO.

Note: In this scenario, the truck and sports utility vehicle are at risk of colliding because the drivers are unable to see one another approaching the intersection and the stop sign is not visible to the driver of the truck. Both drivers would receive warnings of a potential collision, allowing them to take actions to avoid it.

Figure 1. Example of Vehicle-to-Vehicle Communications and a Warning Scenario.

DOT AND INDUSTRY HAVE DEVELOPED AND PILOTED V2V TECHNOLOGIES, WHICH OFFER POTENTIALLY SIGNIFICANT SAFETY BENEFITS IF BROADLY DEPLOYED

Efforts by DOT and the automobile industry to develop V2V technologies have reached the point at which they have been tested in a 12-month, real world pilot that will conclude in February 2014. These efforts have focused on developing and testing needed components including hardware to send and receive data among vehicles, software applications to analyze data and identify potential collisions, vehicle features that issue warnings to drivers of these potential collisions, and a security system to ensure trust in the data that are being communicated among vehicles. According to DOT, once deployed, V2V technologies have the potential to address—by providing warnings to drivers—76 percent of all potential multi-vehicle crashes involving at least one light-duty vehicle.[26] However, the potential benefits of V2V technologies are dependent upon a number of factors including their deployment levels, how drivers respond to warning messages, and the deployment of other safety technologies that can provide similar benefits.

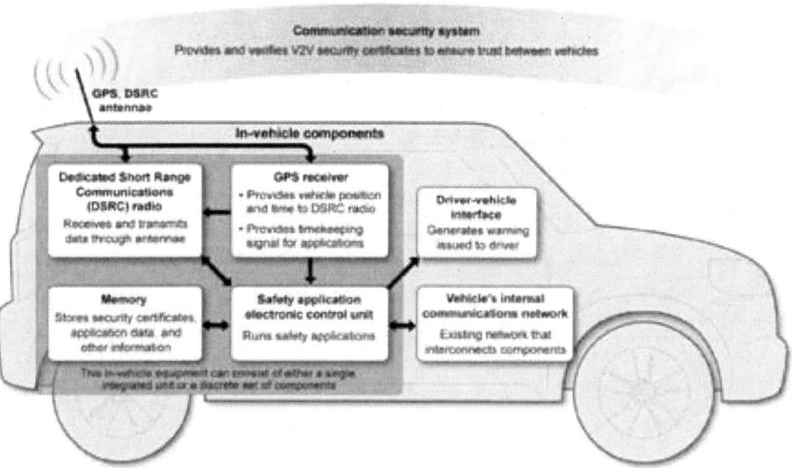

Figure 2. Components of a Vehicle-to-Vehicle Crash Avoidance System.

Development of V2V Technologies Has Evolved to Include a Real World Pilot

Recent and ongoing V2V technology development efforts have been working toward the expected late 2013 NHTSA decision as the next major milestone. DOT, for example, has focused on research and analysis that can provide input for and facilitate its upcoming decision. Most of the 21 experts and 9 automobile manufacturers we spoke with told us that they expect NHTSA's decision to help determine the progress of continued development and eventual deployment of V2V technologies.

Development of these technologies has progressed to the point of real world testing. DOT has sponsored and provided approximately 80 percent of the funding for a pilot test called the Safety Pilot Model Deployment (Safety Pilot). This project—in which DOT has partnered with the CAMP VSC 3 Consortium—has been conducted by the University of Michigan Transportation Research Institute and is taking place in Ann Arbor, Michigan, from August 2012 to February 2014. The primary goals of this pilot are to test the effectiveness of V2V technologies in real world situations and to measure their potential benefits. In total, about 2,700 passenger vehicles were equipped with these technologies in order to participate in the Safety Pilot.[27] NHTSA plans to release findings from the Safety Pilot in the fall of 2014. DOT is considering the data from the first 6 months of the pilot, along with other information, in working toward a decision by late 2013 on how to proceed with V2V technologies.[28]

Efforts by DOT and the automobile industry to develop V2V technologies have focused on both in-vehicle components as well as a security system that manages V2V communications and ensures trust in the data being transmitted among vehicles (see figure 2). Examples of in-vehicle components include:

- V2V hardware, including DSRC radios, cables, and antennae to send and receive data and GPS chips used to determine vehicle location.
- Software applications that analyze data such as location, speed, and brake status and, based on that analysis, predict when collisions are imminent. DOT has worked with automobile manufacturers to determine the most relevant applications based on common crash scenarios and to define, develop, and test these to inform and support NHTSA's V2V technology decision later this year.
- A driver-vehicle interface that—based on the data analysis conducted by the V2V software applications—provides a warning to the driver

through vehicle features such as sounds, lights, or seat vibrations when a collision may be imminent.[29] Existing sensor-based crash avoidance technologies also provide warnings to drivers through similar mechanisms. DOT has been developing guidelines for automobile manufacturers in implementing driver-vehicle interfaces.

In addition to in-vehicle V2V components, an external communication security system is needed to ensure that data being transmitted among vehicles are secure and trusted and have not been altered in the transmission process. Initial research and proposals by DOT and the automobile industry on the design of a communication security system have focused on a public key infrastructure system.[30] Such a system provides security certificates to vehicles which indicate to other vehicles that the data being transmitted are to be trusted because they are valid and have not been altered. Public key infrastructure is used to provide security for many other types of data transmissions and transactions, including online banking and Internet commerce. According to CAMP VSC 3 officials, in-vehicle V2V equipment must be able to detect and automatically report potentially misbehaving devices—such as devices that are malfunctioning, used maliciously, or hacked—to a communication security system. The communication security system must also detect and automatically revoke certificates from vehicles with such devices. Vehicles would receive certificates from a security certificate management authority. However, the technical specifications of how certificates will be provided, how often they will be validated, and who will manage this system have not yet been fully defined. According to DOT officials, a prototype V2V communication security system was developed and tested as part of the Safety Pilot, but that system's design would need additional enhancements to be used for a full deployment.

In-vehicle V2V components could be factory built into new vehicles and fully integrated with existing internal electronics and networks. DOT and the automobile industry have also worked on developing retrofit and aftermarket V2V devices. Such devices would be installed in a vehicle post-manufacture and might not be fully integrated with the vehicle and its existing internal electronics and network. Such aftermarket devices might not have access to all of a vehicle's data, such as brake status or steering wheel angle. As a result, they could provide a less robust set of data to other vehicles and could generate fewer or less precise warning messages to drivers. One specific type of aftermarket device that the automobile industry has worked to develop would only transmit a basic data set about a vehicle's speed and location to other vehicles equipped with V2V technologies; this device would neither receive data from other vehicles nor provide a driver with a warning message, but would interact with the V2V communication security system. DOT is now working with the automobile industry to determine additional standards for such devices to ensure that they work on all types of vehicles and adhere to communication standards to ensure the integrity and security of their data transmission.

Because V2V communications involve the sharing of data among different devices and vehicles of various makes and models, they require technical standards to ensure interoperability. DOT and the automobile industry have worked through international standards organizations such as SAE International and the Institute of Electrical and Electronic Engineers to develop and set standards to facilitate V2V communications. These efforts have focused on standardizing the data elements that vehicles transmit and the means through which data are transmitted.[31]

Although V2V technologies have undergone real-world testing and most experts we interviewed believed that no major technical challenges remain in the development of in-vehicle V2V components, some experts voiced concerns about two technical issues—GPS accuracy and potential channel congestion—issues that DOT and the automobile industry continue to study.

- Three experts we interviewed expressed concern that GPS used by V2V technologies may not provide sufficiently accurate vehicle position for some V2V applications. Two experts also suggested that the means through which V2V technologies use GPS to determine vehicle position (called "relative positioning"[32]) may not be accurate enough, given the need for precise vehicle location data to support V2V software applications. According to DOT officials, however, the department has collected data on the performance of GPS and V2V communications over 20,000 miles in diverse geographic and environmental conditions and is confident that the automotive-grade GPS being used for V2V safety applications is sufficiently accurate. Nevertheless, DOT continues to work to address this issue through research to identify the best means to determine vehicle location through relative positioning. DOT also collected additional data during the Safety Pilot to help ensure that GPS positioning limitations do not negatively affect the performance of V2V technologies.
- Two experts noted that, given the volume of V2V data that would be transmitted in high-traffic areas such as busy highways, channel

congestion on the frequency used by V2V communications could result in delayed transmissions of V2V data. DOT and automobile manufacturers continue to engage in collaborative efforts and sponsor relevant studies to determine the point at which channel congestion might occur and determine potential solutions, such as limiting the frequency of V2V data transmissions in high-traffic situations without compromising safety.[33]

In the longer term, V2V technology development efforts are likely to complement efforts to continue to develop and deploy other sensor-based crash avoidance technologies and autonomous vehicle technologies. NHTSA's Administrator stated in May 2013 that DOT sees these three types of technologies as being part of an evolution from vehicles with limited automatic controls to vehicles with fully autonomous self-driving capabilities. NHTSA's preliminary statement of policy on vehicle automation, issued in May 2013, recognizes the potential benefits of automation and its relationship to V2V technologies.[34] According to DOT officials, DOT's 2015 to 2019 ITS strategic plan, expected to be issued in early 2014, will describe DOT's planned activities in this area. Furthermore, a recent report by an automotive industry consulting firm stated that the convergence of sensor-based crash avoidance technologies and connected vehicle technologies will be needed to enable truly autonomous vehicles, given the benefits and downsides of each type of technology.[35] Five experts and one automobile manufacturer we interviewed said that V2V technologies are a key part of the industry's road map to vehicle automation.

V2V Technologies Have Potential to Provide Significant Safety Benefits Only after Broad Deployment

V2V technologies are expected to offer significant safety benefits by helping drivers avoid collisions in a number of collision scenarios. The automobile industry has worked with DOT to develop safety applications for rear-end, intersection, and lane-change crash scenarios. (See figure 3 for examples of V2V safety applications and a go to http://www.gao.gov/products/GAO-14-13 for a video that demonstrates selected applications warning drivers of potential collisions.) According to NHTSA, these prevalent crash scenarios represent the majority of all vehicle crashes. Because automobiles cannot currently take autonomous actions—such as braking—

based on data gathered by V2V technologies, drivers must act on V2V warnings to prevent collisions from happening. These crash scenario applications are being tested in the Safety Pilot, and DOT and the automobile industry have collected data to determine how to improve the accuracy of the applications. DOT is currently conducting a benefits assessment based in part on the first 6 months of data from the Safety Pilot to inform the decision NHTSA plans to make in late 2013.[36] DOT's Volpe Center[37] is conducting this benefits assessment, which, according to DOT, uses a methodology to determine how many vehicle collisions would be prevented through the use of V2V technologies in a number of different types of collision scenarios. The result will be an estimate of the effectiveness of V2V safety applications.

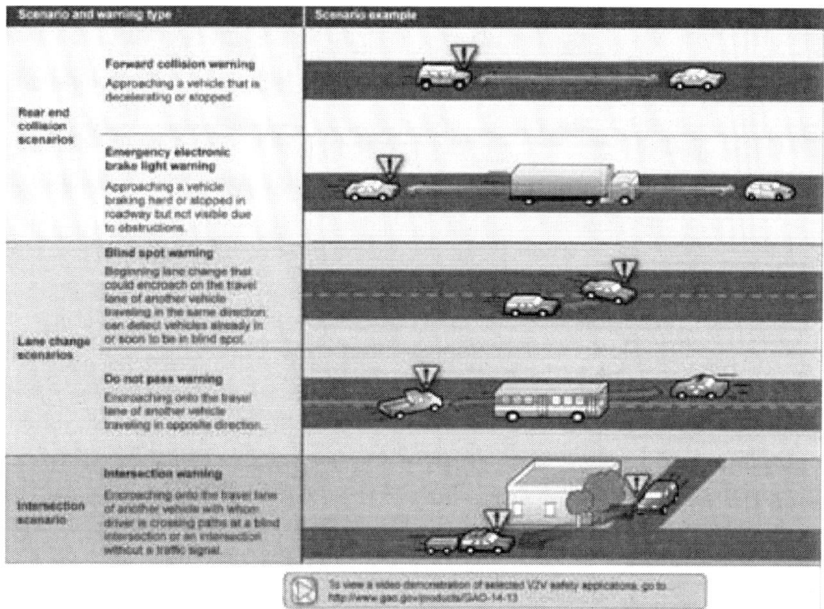

Source: GAO analysis of Crash Avoidance Metrics Partnership information.

Note: Sensor-based crash avoidance technologies can, in some instances, provide warnings in forward collision, blind spot, and do not pass scenarios.

Figure 3. Examples of Crash Scenarios and Vehicle-to-Vehicle Applications.

In May 2013, DOT reported that V2V technologies have the potential to address—by providing warnings to drivers—76 percent of all potential multi-vehicle crashes involving at least one light vehicle.[38] While the majority of experts we interviewed agreed that 76 percent is plausible as a maximum level, this estimate is an upper limit of the potential safety benefits of V2V technologies which assumes their full deployment across the U.S. vehicle fleet. This estimate also assumes that V2V technologies provide warnings in all potential crash scenarios involving at least two vehicles equipped with these technologies. The actual number of crashes prevented by V2V technologies will depend on a number of factors, including the following:

- *Deployment Levels*: According to DOT, the safety benefits of V2V technologies will be maximized with near full deployment across the U.S. vehicle fleet. However, even if NHTSA pursues a rulemaking requiring installation of these technologies in new vehicles, it could take a number of years until benefits are fully realized due to the rate of turnover of the fleet. According to one automobile manufacturer we interviewed, given the rate of new vehicle sales, it can take up to 20 years for the entire U.S. vehicle fleet to turn over. It may be possible to see some benefits at lower levels of deployment, as was seen in the number of interactions among participating vehicles in the Safety Pilot, according to DOT.[39] Also, aftermarket devices that allow existing vehicles to be equipped with V2V devices could help speed deployment. However, three experts we interviewed expressed concern that drivers may not see value in purchasing aftermarket devices, which could limit their adoption.
- *Driver response*: The benefits of V2V technologies will also depend on how well drivers respond to warning messages. If drivers do not take appropriate action in response to warnings, then the benefits of V2V technologies could be reduced. For example, if drivers do not respond to warnings quickly enough—due to distraction, impairment, or other reasons—they may not be able to avoid a collision. Furthermore, if safety applications offer too many false warnings when no imminent threat exists, drivers could begin to ignore valid warnings or not respond to them quickly enough. For example, the Insurance Institute for Highway Safety reported that as many as 41 percent of drivers of certain makes of vehicles with sensor-based lane departure warning systems found the systems "annoying" due to false alarms and unnecessary warnings.[40]
- *Deployment of other safety technologies*: The potential benefits solely attributable to V2V technologies will also depend on the market penetration and effectiveness of sensor-based crash avoidance

technologies. These existing technologies are able to address some of the same crash scenarios as V2V safety applications and their market penetration is likely to increase in the future. While there are cases where V2V technologies can provide safety benefits where sensor-based crash avoidance technologies cannot—such as around a curve or when detecting an unseen stopped car—there are some V2V technology collision scenarios that sensor-based crash avoidance technologies can also address. For example, cameras and radar can be used to provide drivers with forward collision warnings or lane change warnings when another vehicle is in a blind spot. In addition, vehicles in the future may use V2V technologies and sensor- based crash avoidance technologies to complement one another. For example, future vehicles may be able to use V2V data to validate sensor data and provide drivers with more accurate and certain warnings as well as to help execute more reliable automated control actions such as braking. The benefits attributable to V2V technologies will depend on how these technologies work together and the market penetration of these technologies. According to DOT officials, the department's methodology for estimating the safety benefits of V2V technologies will account for the benefits of sensor-based crash avoidance technologies by estimating their penetration and effectiveness in avoiding collisions.

DOT Is Working with the Automobile Industry to Address a Number of V2V Technology Deployment Challenges

While V2V technologies have been tested in real world settings, a number of challenges exist to their wide-scale deployment and the realization of their potential benefits. DOT has been collaborating with automobile manufacturers and others to identify potential solutions to these challenges and is planning continued efforts to support the eventual deployment of V2V technologies. As part of these efforts, the department has been considering various policy options, such as options for managing the V2V communication security system. However, future DOT actions to address some of these issues will not be finalized until after NHTSA decides how the agency will proceed with V2V technologies later this year.

V2V Technology Deployment Challenges and Efforts to Address Them

According to experts we interviewed, DOT officials, automobile manufacturers, and other stakeholders, the deployment of V2V technologies faces a number of challenges, including: (1) finalizing the technical framework and management structure of a V2V communication security system to ensure trust among vehicles, (2) ensuring that the possible sharing of the band of radio-frequency spectrum used by V2V communications will not adversely affect their performance, (3) considering human factors to ensure that drivers respond appropriately to V2V warnings, (4) addressing the uncertainty related to potential liability issues posed by V2V technologies, and (5) addressing any concerns the public may have about V2V technologies, including those related to privacy.[41]

Framework of V2V Communication Security System

As discussed earlier, a security system capable of detecting, reporting, and revoking the credentials of vehicles found to be sharing inaccurate information will be needed to ensure trust in the V2V data transmitted among vehicles. Final plans and policies for the V2V communication security system—including its technical framework and management structure—have not yet been developed and will need to be finalized prior to V2V technology deployment.

- *Technical framework*: Of the 21 experts we interviewed, 12 cited the technical development of a V2V communication security system as a great or very great challenge to the deployment of V2V technologies. One expert told us that it is challenging to establish technical specifications for a system that attempts to maintain users' privacy while providing security for over-the-air transmission of data. Another expert noted that a public key infrastructure system the size of the one needed to support the nationwide deployment of V2V technologies has never been developed before; the sheer magnitude of the system will pose challenges to its development. Both the manner through which V2V communication security certificates will be provided to vehicles and how often they will be provided need to be determined. Some proposals call for using DSRC-equipped roadside infrastructure to provide security certificates, while other proposals call for using non-DSRC communications technologies such as cellular.

- *Management structure*: How a V2V communication security system will be managed has not yet been determined and the delineation of relevant roles and responsibilities could be difficult. Twelve of the experts we interviewed cited the establishment of a management framework for a security system as a great or very great challenge to V2V technology deployment. Because no similar institutions exist, many questions remain about how such a system should be structured and who should manage it. One expert suggested that deciding which approach is most appropriate involves weighing competing philosophies. The expert added that developing the management structure could prove more difficult than developing the technical aspects of the security system.

DOT has researched options to address the technical framework of a V2V communication security system and officials noted that additional technical and policy analyses are under way that will determine the specifics by early 2014. DOT officials stated that a framework that meets the technical requirements of CAMP will be finalized and subsequently prototyped and tested. As mentioned earlier, a prototype technical framework for issuing and managing security certificates has been tested as part of the Safety Pilot. DOT officials told us that there were no significant technical challenges identified during the development of this framework but noted that additional analysis related to the detection and revocation of credentials from vehicles found to be sharing inaccurate information is needed. In preparation for the testing of the Safety Pilot's prototype technical framework, DOT conducted research on the risks associated with such a system and identified potential approaches to address those risks.[42] In addition, the department held public workshops in April 2012 to bring together various stakeholders to discuss issues related to a V2V communication security system.[43] DOT officials explained that NHTSA is currently engaging a firm with expertise in information security systems to perform a review that will help inform the eventual technical specifications of a V2V communication security system. Although various technical frameworks are being considered, limits to DOT's authority could limit the viability of some options. For example, according to DOT officials, the department does not have the legal authority to require the installation of roadside infrastructure to support a V2V communication security system because, aside from roads on federal lands, the federal government does not own or operate U.S. roadways.

In addition, DOT has worked with automobile manufacturers through the VIIC to examine potential management structures for a V2V communication security system. However, officials told us that it would be premature to comment on what the management structure will look like without knowing whether the agency will pursue a rulemaking related to V2V technologies. As part of its research on the technical framework of a V2V communication security system, DOT also sought stakeholder input on the potential management structure and identified three potential options:

- *Federal model*: DOT officials explained that, if the federal government were to provide the security management services required to support V2V technologies, it most likely would do so through a service contract that would include specific provisions to ensure adequate market access, privacy and security controls, and reporting and continuity of services. According to DOT, the department has appropriate legal authority to pursue this model but does not have sufficient resources to introduce such a structure at this time.
- *Public-Private model*: Under a public-private partnership, the security system would be jointly owned and managed by the federal government and private entities. DOT officials stated that statutory authority would be needed to create a public-private management structure that is vested with the authority that Congress deems necessary and appropriate to finance and operate a V2V communication security system. DOT officials added that the required legal authority would likely need to include authorization to establish and collect fees on behalf of the entity and possibly provisions addressing liability, privacy, data ownership, and security requirements applicable to such a system.
- *Private model*: DOT officials stated that its current legal authority and resources have led NHTSA to focus primarily on working with stakeholders to develop a viable private model. DOT officials suggested that, through an agreement with a privately owned and operated security management provider, a private model could be used to support a V2V communication security system. Aside from any aspects specifically detailed in the agreement, DOT suggested that the governance and financing of a private management structure would depend on what entity constitutes and owns the entity.

Potential Spectrum Sharing

In response to requirements in the Middle Class Tax Relief and Job Creation Act of 2012,[44] FCC issued a Notice of Proposed Rulemaking in February 2013 that requested comments on allowing unlicensed devices to share the 5.9 GHz band of the radio-frequency spectrum that had been previously set aside for the use of DSRC-based ITS applications such as V2V technologies.[45] The proposed modifications would provide access to additional spectrum with consistent technical requirements, resulting in faster data speeds by allowing unlicensed devices to use wider bandwidth channels. Existing FCC regulations[46] are designed to ensure that unlicensed devices do not cause interference with licensed users and require operators of unlicensed devices to immediately correct the problem or cease operation if interference occurs.[47] NTIA completed an evaluation in January 2013 in response to requirements in the act that it study spectrum-sharing technologies and the risks to federal users associated with allowing unlicensed devices to share the 5.9 GHz band. NTIA concluded that further work was needed to determine whether and how the risks identified can be mitigated and is currently collaborating with federal and industry stakeholders, as well as FCC, to conduct a quantitative analysis of potential mitigation strategies.[48] FCC officials told us that they are unlikely to act further on a rulemaking in this area until NTIA's analysis is complete.[49]

Although existing FCC regulations are designed to ensure that unlicensed devices do not cause interference, four automobile manufacturers and 16 experts we interviewed expressed concern or uncertainty about the potential effects of allowing unlicensed devices to share the 5.9 GHz band. One automobile industry group said that its members are not opposed to opening the 5.9 GHz band for sharing but emphasized the importance of understanding the implications of doing so to ensure that it will not hinder critical V2V safety applications. As part of the federal rulemaking process, FCC sought public comments on potential approaches to help minimize any potential harmful interference.[50] In response, one automobile manufacturer wrote that it is very concerned that V2V technologies and unlicensed devices will not be able to coexist on the 5.9 GHz band. This automobile manufacturer recommended that FCC proceed with extreme caution when allowing sharing of the 5.9 GHz band, stating that V2V technologies cannot tolerate harmful interference and suggesting that every potential signal loss could render V2V communications ineffective at a moment in which they could protect drivers' lives. Members of the VIIC also noted that the development and testing of V2V technologies has always assumed use of the 5.9 GHz band. Because of this, one expert we interviewed suggested that opening the 5.9 GHz band for sharing would create an added burden for both automobile manufacturers and suppliers, which would have to consider technical steps to make coexistence with unlicensed devices feasible and conduct additional testing to maintain confidence that V2V technologies will work as envisioned. In addition, DOT coordinated with NTIA to submit its concern through comments to FCC that sharing the allocation could degrade the performance of V2V safety applications.

As NTIA continues its analysis of potential risk mitigation strategies, DOT officials told us that the department is working cooperatively with the agency to examine spectrum-sharing arrangements that have been proposed for the 5 GHz band and expects results of this analysis to be available in spring 2014. According to DOT officials, the automobile and Wi-Fi industries are discussing other possible spectrum sharing techniques but specific approaches have not yet been defined.

Human Factors

Because V2V technologies require drivers to take actions based on warning messages,[51] the ultimate effectiveness and safety benefits of V2V technologies depend upon how well drivers respond to the warnings. Addressing the human factors that affect how drivers will respond includes minimizing the risk that drivers could become too familiar with or overly reliant upon warnings over time and fail to exercise due diligence in responding to them, assessing the risk that V2V warnings could distract drivers and present new safety issues, and determining what types of V2V warnings will maximize driver response.

The challenges posed by human factors are in many ways similar to those posed by sensor-based crash avoidance technologies and some other vehicle technologies. However, human factors issues may present even greater challenges to V2V technologies. One automobile manufacturer explained that, since not all vehicles in the United States will be equipped with V2V technologies in the early years of their deployment, it is unknown how drivers will adjust their behavior to account for the fact that not all of the vehicles on the road are capable of providing data. By contrast, with sensor-based technologies, drivers know that their vehicle's warning system is not dependent on the presence of similar technologies in nearby vehicles. Further, the potential introduction of aftermarket V2V devices with a lower level of integration with a vehicle's existing internal network could create additional human factors challenges if aftermarket device warning messages and data are less robust than fully integrated systems. For these reasons, one automobile manufacturer said that it is unrealistic to expect aftermarket devices to perform in all situations.

Automobile manufacturers have already developed different approaches to issuing warnings to drivers for existing sensor-based crash avoidance technologies. Three experts we interviewed suggested that the manner in which V2V warnings are issued to drivers should be allowed to vary among automobile manufacturers once basic standards are put in place. However, two experts suggested that the manner in which V2V warnings are issued to drivers should be similar across vehicles to avoid the confusion that might arise from receiving different types of warnings in similar situations when driving a different vehicle (e.g., a rental car).

NHTSA has a research program in place to develop human factors principles that may be used by automobile manufacturers and suppliers as they design and deploy V2V technology and other safety technology driver-vehicle interfaces that provide warnings to drivers. This program is evaluating alternative approaches to issuing warnings relative to effectiveness and potential driver distraction. In addition, DOT has surveyed drivers participating in the Safety Pilot about driver distraction, in addition to other topics, and collected data on driver responses to various V2V warnings. NHTSA has also sponsored other research on driver warnings—including initial research on the potential standardization of warnings across automobile manufacturers—and is planning to complete a compendium of research findings in late 2013. Based on research in this area, NHTSA has worked with stakeholders to develop design principles for V2V driver-vehicle interfaces, which NHTSA plans to publish in April 2014.[52] DOT continues to determine next steps in this area.[53]

Liability Issues

Six automobile manufacturers and 17 experts we interviewed expressed concern about the challenge posed by uncertainties related to potential liability in the event of a collision involving vehicles equipped with V2V technologies. This challenge is manifested in a number of potential liability issues and questions that are unanswered at this time:

- One automobile manufacturer said that because V2V technologies offer warnings that are based in part on data transmitted by other vehicles—as opposed to sensor-based systems that collect data solely from a vehicle's surroundings—it could be harder to determine whether fault for a collision between vehicles equipped with V2V technologies lies with one of the drivers, an automobile manufacturer, the manufacturer of a V2V device, or another party.

- One expert suggested that, because V2V data may be needed to determine fault in the event of a collision, there could be challenges in determining who owns the data transmitted between vehicles. This expert suggested that establishing rules regarding V2V data ownership would, in the event of a collision, help to answer questions about which parties have access to the data.
- There may be challenges in determining liability if V2V technologies do not work appropriately—for example, if data transmission is delayed due to channel congestion, hacking into the system, or inaccurate GPS readings.
- Four automobile manufacturers shared their concern that the introduction of aftermarket V2V devices into vehicles already on the road would create additional liability issues. One automobile manufacturer explained that this is because it is difficult to integrate aftermarket devices into a vehicle's existing internal network, potentially making it more difficult to gather and transmit the same degree of information as a fully integrated device and, therefore, making it more difficult to determine the cause of a collision.

Automobile manufacturers may be reluctant to move forward with plans to install V2V technologies in their newly manufactured vehicles because of the uncertainty that accompanies these liability issues. Citing some automobile manufacturers' concerns about the liability risks posed by V2V technologies, members of the VIIC have suggested that additional work needs to be done to estimate the potential liability and risks associated with the deployment of V2V technologies and determine ways to mitigate that risk. One automobile manufacturer and two experts suggested that congressional action could be needed to limit the liability levels of automobile manufacturers in the event of a V2V device malfunction. One of these experts suggested that legislation setting forth liability limits for V2V devices that are shown to meet certification tests and function according to regulations and standards may be appropriate. However, one expert suggested that all vehicle technologies involve liability issues and that if the automobile industry ensures that V2V technologies work properly before deployment, V2V technologies should not pose any greater liability risks than existing sensor-based crash avoidance technologies.

DOT officials told us that they do not believe that V2V technologies pose any greater liability issues for automobile manufacturers than existing sensor-based crash avoidance technologies and therefore do not believe that related legislation is necessary. One expert we interviewed suggested that DOT should help guide the process of determining who or what entity owns the data transmitted between vehicles by V2V technologies as that knowledge would make it easier to determine liability in the event of a crash. Another expert suggested that DOT will have a role to play in helping to address liability issues once the specifics of the future deployment of these technologies become clearer.

Public Acceptance

DOT and the CAMP VSC 3 Consortium conducted Driver Acceptance Clinics in 2011 and 2012 to obtain volunteer drivers' feedback on V2V technologies, among other purposes, and participants rated the desirability and usefulness of these technologies highly.[54] However, according to stakeholders and experts we interviewed, obtaining broad public acceptance could present a challenge to deployment, as concerns related to driver privacy and the limited potential benefits of V2V technologies in the early years of their deployment could impede overall public acceptance of these technologies.

- *Privacy concerns*: Public interest groups we interviewed said that overcoming concerns about privacy under a system that involves the sharing of data among vehicles will pose a challenge. One group suggested that the possibility that V2V data could be obtained by third parties such as law enforcement agencies could harm the deployment of these technologies. Similarly, one expert suggested that public acceptance of V2V technologies might be limited without rules prohibiting the use of vehicles' speed and location data to issue tickets or track drivers' movements. Three experts we interviewed suggested that legislation may be needed to limit the potential use of V2V data. Representatives of automobile manufacturers that are members of the VIIC stated that, although the security system under development is being designed to ensure data privacy through a structure that prevents the association of a vehicle's V2V communication security certificates with any unique identifier of drivers of their vehicles, the potential perception of a lack of privacy is a challenge. Further, one automobile manufacturer that is part of the VIIC said that it could be difficult to explain how V2V technologies work to the public without raising concerns related to privacy.[55]

- *Perception of functionality*: Public acceptance of V2V technologies could also be negatively affected by drivers' perception of the technologies' limited functionality, especially when few vehicles are equipped with V2V devices in early years. As mentioned earlier, when there are few equipped vehicles on the roads, drivers may see limited benefits of V2V technologies as they may not frequently encounter similarly equipped vehicles and infrequently receive warning messages in potential collision scenarios. Some experts said that limited benefits in the early years of deployment might pose challenges to consumer acceptance of V2V technologies. For this reason, six experts we interviewed suggested that the simultaneous introduction of applications with mobility applications would help provide benefits to drivers when deployment is low. For example, one expert suggested that mobility applications capable of helping drivers avoid congested roads or notifying them of a broken-down car ahead in the roadway would provide more apparent benefits to drivers, even with a low rate of deployment of V2V technologies.[56] In addition, seven experts we interviewed suggested that it would be helpful for either DOT or automobile manufacturers to reach out to customers to communicate the potential safety benefits of V2V technologies in advance of deployment in order to increase public acceptance of these technologies.

DOT officials told us that the department recognizes that public acceptance needs to be considered for the deployment of V2V technologies and noted that they have worked closely with automobile manufacturers and other stakeholders to develop a technical approach that limits risks to individual privacy. DOT officials have emphasized the need to distinguish between the ability to identify bad actors through a V2V communication security system and the ability to monitor the movements of individual vehicles. DOT stated that as currently conceived, a V2V communication security system would contain multiple technical, physical, and organizational controls to minimize privacy risks—including the risk of vehicle tracking by individuals and government or commercial entities. According to DOT officials, after NHTSA decides whether to proceed with a rulemaking and it is known what entity will provide the security for V2V technologies, the department will perform a comprehensive Privacy Impact Assessment, as required by law,[57] to determine how to balance individual privacy, data security, and safety. DOT officials have emphasized that, to ensure transparency, it will be important to communicate what V2V data is generated by a vehicle, the extent to which it can be linked to drivers, and who or what entities—both legally and technologically—will be able to collect, use, and share the data. DOT officials told us that the department will continue to assess any risks to privacy posed by the introduction of V2V technologies and identify mitigation measures to minimize those risks as more aspects of a system of V2V communications are defined. DOT officials also said that, since its current authority to regulate the interception and use of V2V data is limited, the department might express a strong policy preference or make recommendations to Congress for limitations on the use of V2V data by entities over which the department lacks regulatory authority.

DOT officials told us that the department does not currently have a specific plan about how to proceed with public outreach efforts because NHTSA has not yet decided whether a rulemaking is appropriate at this time. Additional DOT efforts to address other challenges facing the deployment of V2V technologies could help increase their public acceptance. For example, DOT is collecting data from the Safety Pilot regarding the acceptance of V2V technologies by participating drivers. Furthermore, DOT's ongoing development of guidelines for the implementation of driver-vehicle interfaces to address human factors challenges could help increase overall acceptance of these technologies. In addition, DOT has plans for further research on vehicle-to- infrastructure technologies, which could provide additional safety benefits and enable the types of mobility applications that some experts suggested could help increase public acceptance of V2V technologies. Further, DOT is working with industry stakeholders to develop clear certification standards for aftermarket devices, along with all other V2V devices, and is considering incentives to encourage drivers to purchase these devices.

Status of DOT Planning for V2V Technology Deployment

DOT has indicated that its 2015 to 2019 ITS strategic plan—currently under development and expected to be issued in early 2014—will focus on the challenges facing the deployment of V2V technologies. As noted earlier, DOT officials told us that they do not want to take certain actions, such as determining the structure of a V2V communication security system, until after NHTSA's late 2013 decision on how to proceed regarding V2V technologies.

DOT has stated that the department's forthcoming plan will include details of near term efforts to further develop the technical framework and management structure of a V2V communication security system and consideration of incentives needed to support the eventual deployment of V2V technologies to the U.S. vehicle fleet. In addition, DOT plans to develop an understanding of how the public views connected vehicle technologies and their benefits in order to develop effective methods for describing the value of V2V technologies to users in the context of partial deployment.

According to DOT, the strategic vision for the 2015 to 2019 ITS strategic plan is to improve safety, mobility, and environmental mitigation through connected vehicles and to enable a transportation system that builds upon the capabilities of V2V technologies and vehicle-to-infrastructure technologies. In line with this goal, as noted earlier, DOT is planning for additional research, including research conducted with the automobile industry, into vehicle-to-infrastructure technologies. In addition, by defining how an automated vehicle fleet can be introduced with limited impact to current infrastructure and researching how increases in the degree of automation of the U.S. vehicle fleet could have liability implications for stakeholders, DOT has stated that it plans to prepare for the introduction of autonomous vehicles to a connected vehicle environment featuring both V2V and vehicle-to-infrastructure technologies.

Costs of V2V Technologies Are Being Studied and Are Likely to Be Influenced by Various Factors Including Specifics of V2V Communication Security System

DOT and the automobile industry, through the CAMP VSC 3 Consortium, are currently analyzing the total costs of deploying V2V technologies, which include the costs of in-vehicle components and the costs associated with a V2V communication security system. DOT is currently obtaining estimates of the costs of in-vehicle V2V components from automobile manufacturers and industry suppliers and has engaged a contractor to study the potential costs of providing V2V communications security through a number of possible technical and management options. NHTSA will use these cost estimates to inform its decision on how to proceed with V2V technologies later this year. In addition, the CAMP VSC 3 Consortium is now conducting an independent analysis of potential security system costs. According to an official with that organization, this study should be completed in late 2013 and the consortium will provide input and comments on the key assumptions of DOT's cost study as well. Despite these efforts, all of the automobile manufacturers we interviewed told us that they had not yet completed any formal studies estimating V2V technology costs. We also conducted a literature search for published studies discussing potential V2V technology costs and were unable to identify any such studies. Finally, we asked the experts we interviewed to identify studies on the potential costs of deploying V2V technologies; none of the 21 experts were able to identify any studies beyond those now being conducted by DOT or the CAMP VSC 3 Consortium.

All of the automobile manufacturers we interviewed said that it is difficult to estimate the costs of in-vehicle V2V components at this time because too many factors remain unknown. According to both automobile manufacturers and experts we interviewed, a number of factors will influence the costs of in-vehicle V2V components:

- *Volume*: Five experts and two automobile manufacturers we interviewed pointed out that as the volume of V2V components produced increases, the per-unit costs are likely to decrease due to the economies of scale in manufacturing. Thus, a federal requirement that automobile manufacturers install V2V technologies in newly manufactured vehicles would likely result in lower per-vehicle costs of in-vehicle V2V components than would otherwise be the case.
- *Time frames*: The costs of in-vehicle components of V2V technologies may decline over time due to technological advances. Two automobile manufacturers we interviewed said that both the automobile and consumer electronics industries have been successful at driving down the costs of other technologies over time and suggested that this also may occur with the cost of in-vehicle V2V components.
- *Degree of integration with existing vehicle technologies*: Automobile manufacturers already install sensor-based crash avoidance technologies in some vehicle models. The extent to which V2V technologies are integrated with the existing components of these and other technologies that use similar components will influence the costs of V2V technologies. For example, vehicles that already have GPS chips installed for navigation purposes may not need an additional GPS chip specifically for V2V technologies purposes. One expert and one automobile manufacturer we interviewed explained

that automobile companies have been successful in integrating different technologies in their vehicles in recent years, resulting in reduced costs.

- *Federal requirements*: The specifics of any federal regulation regarding V2V technologies, including any possible regulation mandating their installation in new vehicles, would likely influence costs. For example, the introduction of requirements related to V2V technological specifications and standards, the use of hardware for V2V components, and the use of V2V safety applications would influence the ultimate costs of V2V technologies.

Although the costs of in-vehicle V2V components are difficult to estimate at this time, they may be modest compared to the price of a new vehicle. According to CAMP VSC 3 Consortium representatives, in-vehicle V2V components—including the DSRC components and GPS receiver—are already commercially available at a relatively low cost. A CAMP VSC 3 representative noted that although members of the partnership are unable to discuss potential V2V technology costs with one another due to antitrust concerns, they have agreed to focus their V2V technology development efforts on limiting their costs as much as possible. As a result, compared to the average sale price of a new vehicle—about $31,000 in 2012, according to the National Automobile Dealers Association—the potential costs of these physical components may not add significantly to a vehicle's price. One expert we interviewed estimated that the costs of V2V components would add less than one percent to the price of a new vehicle. These costs may be less than the costs of sensor- based crash avoidance technologies, according to one industry association we interviewed, because sensors are more expensive.

At this time, the potential costs associated with a V2V communication security system are unknown as specifics of a security system remain undetermined. One expert we interviewed explained that since it is not yet known how the security system will look or how it will be organized or managed, no one has a good handle on those costs. According to DOT officials, how often vehicles must communicate with the security system and requirements of how security certificates will be provided will influence the ultimate cost of a V2V communication security system. Furthermore, it is currently not only difficult to estimate the potential costs, but unclear who or what entity—consumers, automobile manufacturers, DOT, state and local governments, or others—would pay the costs. Determining who or what entity will fund the system will likely prove challenging.

Although the potential costs of a V2V communication security system are unknown at this time, eight experts we interviewed noted that either these costs, or the costs of roadside equipment that may be needed, could be significant. One expert said that, given the need for continuous operation of a security system, the costs are likely to be much greater than the costs of in-vehicle V2V components. Potential sources of funding may also face challenges in providing it. For example, if roadside equipment is needed to support a security system, it could be costly in a nationwide deployment. Participants in a 2012 DOT-hosted workshop focused on connected vehicle security[58] noted that the costs of roadside equipment could range between $25,000 and $30,000 per installation, a cost that many state and local governments would find prohibitive if they were responsible for its financing.[59] In addition, three experts whom we interviewed said that the costs of ongoing maintenance and operation of such installations could also prove prohibitive. According to DOT officials, however, the potential costs of roadside equipment could be much lower if its installation were to be integrated into already planned construction work at selected sites. They also noted that these costs should decline over time due to declining costs of technologies and increased volumes of production.

Agency Comments

We provided a draft of this report to the Secretary of Transportation and the Chairman of the Federal Communications Commission for review and comment. DOT and FCC both provided comments via email that were technical in nature. We incorporated these comments as appropriate.

David J. Wise
Director, Physical Infrastructure Issues

Appendix I: Scope and Methodology

To address all of our objectives, we reviewed documentation of the efforts of the U.S. Department of Transportation (DOT) and automobile manufacturers related to vehicle-to-vehicle (V2V) technologies, such as the department's ITS Strategic Research Plan, 2010 - 2014 Progress Update 2012 and documentation on completed and ongoing research. We also interviewed officials from DOT's National Highway Traffic Safety Administration (NHTSA) and the Research and Innovative Technology Administration (RITA) about these efforts. We also interviewed officials with the Crash Avoidance Metrics Partnership (CAMP) Vehicle Safety Communications (VSC) 3 Consortium as well as, both collectively and individually, representatives from all of the following automobile manufacturers that currently comprise the Vehicle Infrastructure Integration Consortium: BMW, Chrysler, Ford,[60] General Motors, Honda, Hyundai-Kia, Mercedes-Benz, Nissan, Toyota, and Volkswagen. We also interviewed a V2V device supplier and representatives of industry and public interest groups knowledgeable on the topic of V2V technologies, such as the Alliance of Automobile Manufacturers and the American Automobile Association.

In addition, we collaborated with the National Academies of Sciences (NAS) to identify and recruit experts in vehicle-to-vehicle technologies. We provided NAS with criteria for selecting experts, which included: (1) type and depth of experience, including recognition in the professional community and relevance of any published work; (2) employment history and professional affiliations, including any potential conflicts of interest; and (3) other relevant experts' recommendations. NAS initially identified 35 experts knowledgeable in the areas of V2V technology development and interoperability, technology deployment, production of light-duty passenger vehicles, data privacy and security, legal and policy issues, and human factors issues related to V2V technologies. From that list, we selected and conducted structured interviews with 21 experts who represented domestic and international automobile manufacturers, suppliers of V2V devices, a telecommunications company, and state governments, as well as automotive industry experts and academic researchers (see table 1 for list of experts interviewed). In conducting our structured interviews, we used a standardized interview guide (see appendix II) to obtain consistent answers. During these interviews we asked, among other things, for expert views on the state of development of V2V technologies, the potential benefits of V2V technologies, their potential costs, and DOT's role in developing V2V technologies. We also asked for each expert's views on a number of already defined potential challenges facing the deployment of V2V technologies. We determined this initial list of potential challenges after initial interviews with DOT, industry associations, select automobile manufacturers, and other interest groups knowledgeable about V2V technologies. Prior to conducting the interviews, we pretested the structured interview guide with two of the selected experts to ensure our questions were worded appropriately and could be administered consistently. After conducting these interviews, we conducted a content analysis of expert responses relevant to each objective and counted the rating of each potential challenge discussed. See appendix III for a count of the ratings provided by experts on potential challenges facing deployment of V2V technologies.

Table 1. Subject Matter Experts Interviewed

Name	Affiliation[a]
Roger Berg	DENSO
Chris Body	Kapsch TrafficCom
John Campbell	Battelle
Susan Chrysler	University of Iowa
Richard Deering	Richard Deering Associates
Thomas Dingus	Virginia Tech
Dorothy Glancy	Santa Clara University School of Law
John Kenney	Toyota
Bob Koeberlein	Idaho Transportation Department
Greg Krueger	Science Applications International Corporation (SAIC)
Mike Manser	University of Minnesota
Jim Misener	Independent consultant
Ravi Puvvala	Savari
Bob Rausch	Transcore
Russell Shields	Ygomi
Steve Shladover	California Partners for Advanced Transportation Technology (PATH)
Mike Shulman	Ford Motor Company
Kirk Steudle	Michigan Department of Transportation
Mitch Tseng	Tseng Infoserv
Bryant Walker Smith	Stanford Law School
William Whyte	Security Innovation

Source: GAO.

[a] We interviewed experts as individuals, not as representatives of any institution. We provide information on institutions to help readers identify experts.

We completed a literature search to obtain documentation, studies, and articles related to our objectives. Although this report focuses upon V2V technologies, our literature search was broadened to include any relevant work published in the past 10 years that was related to terms including "vehicle-to-vehicle communications," "vehicle-to-infrastructure communications," "vehicle-to-roadside communications," or "intelligent vehicle highway systems."

To specifically address the state of development of V2V technologies and their anticipated benefits, we conducted a site visit to Ann Arbor, Michigan, where we interviewed researchers from the University of Michigan Transportation Research Institute who are managing the Safety Pilot Model Deployment and toured areas of the city in which related infrastructure was installed. We also interviewed automobile manufacturers' CAMP VSC 3 Consortium representatives in Farmington Hills, Michigan, and received a demonstration of V2V safety warnings in multiple potential crash scenarios. We also reviewed DOT's May 2013 Description of Light-Vehicle Pre-Crash Scenarios for Safety Applications Based on Vehicle-to-Vehicle Communications report that estimates potential benefits of V2V technologies.

In addition, to specifically address the challenges facing the deployment of V2V technologies, we interviewed officials from the Federal Communications Commission and a privacy group to obtain their views on the potential challenges of spectrum allocation and data privacy, respectively, related to deployment of V2V technologies.

To specifically address the potential costs associated with V2V technologies, we discussed efforts to estimate costs with DOT and representatives of the CAMP VSC 3 Consortium and reviewed relevant documentation, such as the April 2012 public workshop entitled Enabling a Secure Environment for Vehicle-to-Vehicle (V2V) and Vehicle-to-Infrastructure (V2I) Transactions, which DOT organized to facilitate discussion of communications security structures under development among public and private stakeholders. In addition, we asked all automobile manufacturers we interviewed about the potential costs of V2V technologies.

We conducted this performance audit from October 2012 through November 2013 in accordance with generally accepted government auditing standards. Those standards require that we plan and perform the audit to obtain sufficient, appropriate evidence to provide a reasonable basis for our findings and conclusions based on our audit objectives. We believe that the evidence obtained provides a reasonable basis for our findings and conclusions based on our audit objectives.

APPENDIX II: STRUCTURED INTERVIEW GUIDE FOR EXPERTS IDENTIFIED BY THE NATIONAL ACADEMIES OF SCIENCES

Questions for Connected Vehicle Technology Experts

(Experts identified by the Transportation Research Board of the National Academies)

Overview of GAO Review of Connected Vehicle Technologies

The United States Government Accountability Office (GAO) is undertaking work looking at the use of connected vehicle technologies (CVT) involving vehicle-to-vehicle (V2V) communications. These technologies allow vehicles to wirelessly communicate with one another through Dedicated Short Range Communications (DSRC); share data, such as data on vehicle location and speed; and provide drivers with warnings of potential collisions. GAO is examining these technologies at the request of the House Committee on Science, Space, and Technology and Committee on Transportation and Infrastructure, Subcommittee on Highways and Transit. Specifically, we are examining:

(1) the progress of development of connected vehicle technologies that involve vehicle-to-vehicle communications and their anticipated benefits;
(2) the challenges that affect the development and deployment of these technologies and their potential costs; and
(3) how the U.S. Department of Transportation is leading efforts to address these challenges.

Our work is limited to CVTs involving V2V communications and we are not including vehicle-to-infrastructure (V2I) communications beyond the extent to which infrastructure may be needed to support V2V communications. We are reviewing efforts in the United States and not in other countries. In addition, our work is limited to passenger vehicles and we are not including commercial and transit uses.

Please keep the following in mind as you consider your responses to the questions:

- We use the term "stakeholders" to indicate organizations that have been or are expected to be involved in the development or deployment of CVTs involving V2V communications including, but not limited to, automobile manufacturers, parts suppliers, government agencies, and telecommunications companies.
- We use the term "CVTs involving V2V communications" to include devices that send and receive data, safety applications (such as Forward Collision Warning and Emergency Electronic Brake Lights Warning) that analyze data and provide warnings through driver-vehicle interfaces, and any security system(s) that may be needed to help establish trust between vehicles sharing data.
- We use the term "V2V devices" to represent the devices now being tested in the Safety Pilot Model Deployment in Ann Arbor, Michigan:[61]
  1) ***Fully-integrated devices:*** these devices would be installed during vehicle production and send, receive and analyze vehicle data to generate driver warnings of potential collisions. Because they are fully integrated into vehicles, they can analyze additional vehicle data—such as steering wheel angle, acceleration rate, brake status and turn signal status—to provide more robust warnings to drivers and provide more complete data to other vehicles.
  2) ***Aftermarket safety devices:*** these devices, installed after vehicle production, send, receive, and analyze vehicle data to generate warnings of potential collisions. However, because they are not integrated with vehicles' electronics architecture and are thus unable to consider additional data, these devices may be limited in their warnings and driver-vehicle interface.
  3) ***Vehicle awareness devices:*** these devices are installed after vehicle production and are not connected to any vehicle systems. They send data on a vehicle's location and speed which may be received by other vehicles, but these devices do not generate driver warnings.

Status and Benefits of Vehicle-to-Vehicle Connected Vehicle Technologies

1) Please discuss your views of the status of development of CVTs that involve V2V communications in the United States, including progress made and successes to date.

2) In your opinion, what are some of the greatest potential benefits of CVTs that involve V2V communications?
   a) Which potential V2V applications are likely to offer the greatest benefits and why?
3) What work, in your opinion, needs to be completed before such technologies are commercially feasible and available in the U.S.?
   a) In your opinion, what is a realistic timeframe in which automobile manufacturers might begin to install these technologies in new vehicles being manufactured for the U.S. vehicle fleet?
4) According to a 2007 report by the U.S. DOT, CVTs involving V2V communications have the potential to address up to 76 percent of unimpaired roadway crashes, which could have the potential to greatly reduce the number of roadway fatalities that occur each year.
   a) What views, if any, do you have on DOT's projection of these technologies' potential benefits?
   b) Other than this study and DOT's Safety Pilot Driver Clinics and Model Deployment, are you aware of any other studies that estimate the potential benefits of V2V technologies?

Costs of Vehicle-to-Vehicle Connected Vehicle Technologies

5) To your knowledge, are there any studies that estimate the potential future costs associated with the deployment of CVTs involving V2V communications?
   a) Specifically, are you aware of any work that has been done to estimate the costs associated with the deployment of (1) fully-integrated devices, (2) aftermarket safety devices, and (3) vehicle awareness devices?
6) To your knowledge, are there any studies that estimate the potential costs of the type(s) of security system(s) needed to support CVTs involving V2V communications, including any infrastructure that such a security system may require?
7) What factors may affect the costs associated with the deployment of CVTs involving V2V communications?

Challenges Associated with Vehicle-to-Vehicle Connected Vehicle Technologies

8) In your opinion, to what extent does each of the following issues present a challenge to the development and eventual deployment of

CVTs that involve V2V communications? *(Place only one 'x' in each row below to indicate your response.)*

	Very great challenge	Great challenge	Moderate challenge	Slight challenge	No challenge	Don't know
a. Technical challenges in the development of each of the following V2V technologies:						
- V2V devices	☐	☐	☐	☐	☐	☐
- V2V safety applications	☐	☐	☐	☐	☐	☐
- Driver-vehicle interface	☐	☐	☐	☐	☐	☐

i. Please discuss the technical challenges being faced in the development of CVTs involving V2V communications.

ii. Does this issue present different challenges for different relevant stakeholders? If so, please discuss.

iii. Please discuss how well U.S. DOT and other stakeholders are addressing this challenge and any additional steps they should take.

	Very great challenge	Great challenge	Moderate challenge	Slight challenge	No challenge	Don't know
b. Technical development of a data security system to help establish trust between vehicles sharing information	☐	☐	☐	☐	☐	☐

i. Please discuss any challenges posed by issues of technical development of a data security system to the development and deployment of CVTs involving V2V communications.

ii. Does this issue present different challenges for different relevant stakeholders? If so, please discuss.

iii. Please discuss how well U.S. DOT and other stakeholders are addressing this challenge and any additional steps they should take.

	Very great challenge	Great challenge	Moderate challenge	Slight challenge	No challenge	Don't know
c. Standardization to ensure interoperability among different types of devices and vehicles	☐	☐	☐	☐	☐	☐

i. Please describe the challenge that ensuring the interoperability of CVTs involving V2V communications presents to development and deployment of these technologies.
ii. Does this issue present different challenges for different relevant stakeholders? If so, please discuss.
iii. Please discuss how well U.S. DOT and other stakeholders are addressing this challenge and any additional steps they should take.

	Very great challenge	Great challenge	Moderate challenge	Slight challenge	No challenge	Don't know
d. Deployment of V2V devices and applications into a great enough percentage of the U.S. vehicle fleet to realize significant benefits	☐	☐	☐	☐	☐	☐

i. Please discuss the challenge(s) associated with the process of deploying V2V technologies across the U.S. vehicle fleet.
   a. Please discuss any challenges presented by the potential need for less-than fully integrated aftermarket V2V devices.
ii. Does this issue present different challenges for different relevant stakeholders? If so, please discuss.
iii. Please discuss how well U.S. DOT and other stakeholders are addressing this challenge and any additional steps they should take.

	Very great challenge	Great challenge	Moderate challenge	Slight challenge	No challenge	Don't know
e. Costs of deploying CVTs involving V2V communications (including V2V devices, safety applications, driver-vehicle interface, and security system)	☐	☐	☐	☐	☐	☐

i. Please discuss any challenge(s) posed by the costs associated with specific aspects of the deployment of CVTs involving V2V communications.
ii. Does this issue present different challenges for different relevant stakeholders? If so, please discuss.
iii. Please discuss how well U.S. DOT and other stakeholders are addressing this challenge and any additional steps they should take.

	Very great challenge	Great challenge	Moderate challenge	Slight challenge	No challenge	Don't know
f. Potential need for roadside equipment required to support security system	☐	☐	☐	☐	☐	☐

i. Please discuss the extent to which the potential need for roadside equipment to support V2V technologies may or may not present a challenge(s).
ii. Please discuss the extent to which financing any roadside equipment needed for a system of V2V technologies may or may not present a challenge(s).
iii. Does this issue present different challenges for different relevant stakeholders? If so, please discuss.
iv. Please discuss how well U.S. DOT and other stakeholders are addressing this challenge and any additional steps they should take.

	Very great challenge	Great challenge	Moderate challenge	Slight challenge	No challenge	Don't know
g. Establishment of a framework, including determination of roles and responsibilities, for management of a security system to support a V2V environment	☐	☐	☐	☐	☐	☐

i. Please discuss the challenge(s) associated with the establishment of a governance structure for the management of a security system to support CVTs involving V2V communications.
ii. Does this issue present different challenges for different relevant stakeholders? If so, please discuss.
iii. Please discuss how U.S. DOT and other stakeholders are addressing this challenge and any additional steps they should take.

	Very great challenge	Great challenge	Moderate challenge	Slight challenge	No challenge	Don't know
h. Establishing acceptable end user privacy	☐	☐	☐	☐	☐	☐

i. Please discuss any challenges posed by issues of end user privacy to the development and/or deployment of V2V technologies.

ii. Does this issue present different challenges for different relevant stakeholders? If so, please discuss.

iii. Please discuss how well U.S. DOT and other stakeholders are addressing this challenge and any additional steps they should take.

	Very great challenge	Great challenge	Moderate challenge	Slight challenge	No challenge	Don't know
i. Public acceptance	☐	☐	☐	☐	☐	☐

i. Please discuss the challenge of public acceptance to the development and deployment of V2V technologies.

ii. Does this issue present different challenges for different relevant stakeholders? If so, please discuss.

iii. Please discuss how well U.S. DOT and other stakeholders are addressing this challenge and any additional steps they should take.

	Very great challenge	Great challenge	Moderate challenge	Slight challenge	No challenge	Don't know
j. Use of DSRC technology at 5.9 GHz frequency band	☐	☐	☐	☐	☐	☐

i. Please discuss any challenge(s) presented by the decision to use DSRC technology to establish V2V communications to date and in the future.

ii. The Federal Communications Commission allocated 75 MHz of spectrum in the 5.9 GHz band for vehicle safety and mobility applications. Please discuss any implications of the possibility of making the bandwidth set aside for CVTs for other uses.

iii. Does this issue present different challenges for different relevant stakeholders? If so, please discuss.

iv. Please discuss how well U.S. DOT and other stakeholders are addressing this challenge and any additional steps they should take.

	Very great challenge	Great challenge	Moderate challenge	Slight challenge	No challenge	Don't know
k. Human factors for V2V communications *(e.g. how warning messages affect driver performance)*	☐	☐	☐	☐	☐	☐

i. Please discuss the challenges that human factors pose to the development and eventual deployment of CVTs that involve V2V communications.
ii. Does this issue present different challenges for different relevant stakeholders? If so, please discuss.
iii. Please discuss how well U.S. DOT and other stakeholders are addressing this challenge and any additional steps they should take.

	Very great challenge	Great challenge	Moderate challenge	Slight challenge	No challenge	Don't know
1. Liability issues *(i.e. uncertainty related to legal responsibility for vehicle crashes after broad deployment of V2V devices)*	☐	☐	☐	☐	☐	☐

i. Please discuss the extent to which the broad introduction of V2V technologies to the U.S. vehicle fleet may pose potential challenges related to liability.
ii. Does this issue present different challenges for different relevant stakeholders? If so, please discuss.
iii. Please discuss how well U.S. DOT and other stakeholders are addressing this challenge and any additional steps they should take.

9) What additional challenges, if any, exist in the ongoing development and eventual deployment of CVTs that involve V2V communications? For each additional challenge, please discuss its extent, how it presents different challenges for different relevant stakeholders, and how well DOT and other stakeholders are addressing it and additional steps that they should take.

	Very great challenge	Great challenge	Moderate challenge	Slight challenge	No challenge	Don't know
a. ______	☐	☐	☐	☐	☐	☐

	Very great challenge	Great challenge	Moderate challenge	Slight challenge	No challenge	Don't know
b. ______	☐	☐	☐	☐	☐	☐

	Very great challenge	Great challenge	Moderate challenge	Slight challenge	No challenge	Don't know
c. ______	☐	☐	☐	☐	☐	☐

U.S. DOT's Role in Connected Vehicle Technologies

10) In your view, how well is U.S. DOT carrying out the following efforts related to the development and deployment of CVTs involving V2V communications?
    a. Evaluating the potential safety benefits of CVTs involving V2V communications
    b. Considering the potential costs of deploying CVTs involving V2V communications
    c. Supporting the development of V2V safety applications.
    d. Conducting development and testing of V2V devices and promoting the development of technical standards.
    e. Communicating and collaborating with key stakeholders, including automobile manufacturers and suppliers and other interested parties.
    f. Identifying policies to help lead to deployment of CVTs involving V2V communications.
11) Please discuss how, in your opinion, each of the following possible actions regarding CVTs involving V2V communications would affect automobile manufacturer costs, new vehicle costs, vehicle production, and deployment of CVTs involving V2V communications:
    a. A federal mandate for automobile manufacturers to install equipment necessary to support V2V safety applications?

b. The inclusion of V2V safety applications in NHTSA's New Car Assessment Program?[62]
c. Continued research and development?

12) Do you have any views not already discussed on the development and deployment of CVTs involving V2V communications, including any additional actions U.S. DOT should take?

APPENDIX III: EXPERT RATINGS OF POTENTIAL CHALLENGES FACING DEPLOYMENT OF VEHICLE-TO-VEHICLE TECHNOLOGIES

As part of our review, we conducted 21 structured interviews with individuals identified by the National Academies of Sciences to be experts on vehicle-to-vehicle (V2V) technologies (see table 1 in appendix I for list of experts interviewed). In conducting these structured interviews, we used a standardized interview guide (see Appendix II) to obtain consistent answers. During these interviews we asked, among other things, for each expert's views on a number of already defined potential challenges facing the deployment of V2V technologies. The ratings provided by the experts for each of the potential challenges discussed are shown in table 2 below. To inform our discussion of the challenges facing the deployment of V2V technologies, we considered these ratings as well as experts' responses to open-ended questions.

Table 2. Expert Ratings of Potential Challenges Facing Deployment of Vehicle-to-Vehicle Technologies

Potential Challenge	Very great challenge	Great challenge	Moderate challenge	Slight challenge	No challenge	Don't know
Technical challenges in the development of each of the following V2V technologies:						
V2V devices	0	2	9	7	2	1
V2V safety applications	0	3	12	4	1	1
Driver-vehicle interface	4	1	9	5	1	1
Technical development of a data security system to help establish trust between vehicles sharing information	6	6	4	3	0	2

Potential Challenge	Very great challenge	Great challenge	Moderate challenge	Slight challenge	No challenge	Don't know
Standardization to ensure interoperability among different types of devices and vehicles	1	6	6	6	2	0
Deployment of V2V devices and applications into a great enough percentage of the U.S. vehicle fleet to realize significant benefits	4	4	8	2	0	3
Costs of deploying connected vehicle technologies involving V2V communications (including V2V devices, safety applications, driver-vehicle interface, and security system)	2	4	4	7	3	1
Potential need for roadside equipment required to support security system	1	8	6	3	0	3
Establishment of a framework, including determination of roles and responsibilities, for management of a security system to support a V2V environment	4	8	6	1	1	1
Establishing acceptable end user privacy	6	1	7	4	1	2
Public acceptance	2	7	7	3	1	1
Use of DSRC technology at 5.9 GHz frequency band	6	4	5	1	0	5
Human factors for V2V communications (e.g. how warning messages affect driver performance)	4	3	6	5	0	3
Liability issues (i.e. uncertainty related to legal responsibility for vehicle crashes after broad deployment of V2V devices)	4	4	5	3	0	5

Source: GAO analysis of structured interviews with experts identified by the National Academies of Sciences.

End Notes

[1] In addition, NHTSA has estimated that motor vehicle crashes had economic costs— including productivity losses, property damages, and medical costs—of $230 billion in 2000 (about $304 billion in 2013 dollars). See L. Blincoe et al., *The Economic Impact of Motor Vehicle Crashes* (Washington, D.C.: NHTSA, 2002). NHTSA has a study under way to update crash cost estimates and expects to issue results by the end of 2013.

[2] 49 USC §§ 30101 and 30111.

[3] In some cases, certain vehicles can take action without driver input, such as emergency braking, based on such warnings. In addition, such technologies can use sensors to alert drivers when a vehicle is moving out of its lane of travel without use of a turn signal.

[4] The automobile industry has also been researching and developing autonomous vehicle technologies. These would allow for vehicle operation without direct driver input.

[5] NHTSA will complement this 2013 decision with a 2014 decision on introducing such technology for commercial vehicles, including trucks and buses. The Federal Highway Administration is developing guidance for release in 2015 to assist state departments of transportation to adapt to these new technologies.

[6] DOT collaborated with a consortium of automobile manufacturers to sponsor the Safety Pilot Model Deployment, a pilot test being conducted by the University of Michigan Transportation Research Institute that is taking place in Ann Arbor, Michigan. The pilot began in August 2012 and is scheduled to end in February 2014.

[7] The term "human factors" refers broadly to how humans' abilities, characteristics, and limitations interact with the design of the equipment they use and the environments in which they function.

[8] However, NHTSA's preliminary data for 2012 indicate that the total number of traffic fatalities increased to approximately 34,000 and the rate of traffic fatalities per 100 million vehicle miles traveled increased to approximately 1.16.

[9] Radio detection and ranging, or RADAR, is a method of detecting objects and determining their position, velocity, or other characteristics by analysis of high frequency radio waves reflected from their surfaces. Light detection and ranging, or LiDAR, uses a narrow beam to transmit infrared light pulses to a target, which then travel back again. Measurement of the elapsed time of the light beam to reach the vehicle and return computes the vehicle's distance from the operator.

[10] In November 2012, the National Transportation Safety Board urged NHTSA to establish performance standards for crash-avoidance technologies and mandate that such technologies be included as standard equipment in motor vehicles. Currently, sensor- based crash avoidance technologies are not required to be installed in new vehicles. In 2010, NHTSA announced plans to mandate the installation of rear-mounted video cameras and in-vehicle displays in all new passenger vehicles by September 2014 but has not yet established rules governing the placement or minimum field of vision of such cameras.

[11] A forward collision warning system can issue a warning to drivers when their vehicle is about to collide with a vehicle in front of them. Lane departure warnings can alert a driver when their vehicle begins to veer outside of a lane of travel without the driver activating a turn signal.

[12] ITS technologies consist of a range of communications, electronics, and computer technologies, such as systems that collect real-time traffic data and transmit information to the public via means such as dynamic message signs, ramp meters to improve the flow of traffic on freeways, and synchronized traffic signals that are adjusted in response to traffic conditions.

[13] DOT's early work on connected vehicle technologies focused on vehicle-to-infrastructure technologies. The department increased its focus on V2V technologies because they are projected to produce the majority of connected vehicle technology safety benefits and they do not require the same level of infrastructure investment as vehicle-to-infrastructure technologies. DOT officials indicated that the department plans to increase its focus on vehicle-to-infrastructure technologies starting in 2015.

[14] Vehicle-to-infrastructure technologies can also provide additional safety benefits. For example, they can provide drivers with information regarding safe speeds and road conditions such as work zones and, for example, warn drivers when they may be approaching a curve at an unsafe speed.

[15] The CAMP VSC 3 Consortium is currently comprised of Ford Motor Company; General Motors Holdings LLC; Honda R&D Americas, Inc.; Hyundai America Technical Center,

Inc.; Mercedes-Benz Research & Development North America, Inc.; Nissan Technical Center North America; Toyota Motor Engineering & Manufacturing North America, Inc.; and Volkswagen Group of America, Inc.

[16] VIIC is currently comprised of BMW of North America, LLC; Chrysler Group LLC; Ford Motor Company; General Motors Holdings LLC; Honda R&D Americas, Inc.; Hyundai America Technical Center, Inc.; Mercedes-Benz Research & Development North America, Inc.; Nissan Technical Center North America; Toyota Motor Engineering & Manufacturing North America, Inc.; and Volkswagen Group of America, Inc.

[17] A number of states have enacted legislation permitting the operation of these vehicles under certain conditions and, according to Google, its fleet of autonomous vehicles had logged more than 300,000 miles as of August 2012. In May 2013, DOT announced plans for research on safety issues related to autonomous vehicles and offered recommendations for states related to the testing, licensing, and regulation of these vehicles.

[18] Vehicles with adaptive cruise control are able to adjust their own speed based on the distance between the vehicle and vehicles ahead in order to maintain a safe distance.

[19] Allocation involves segmenting spectrum, the natural resources used for wireless communication, into bands of frequencies that are designated for use by particular types of services. FCC manages spectrum use for nonfederal users, including commercial, private, and state and local government users; the Department of Commerce's National Telecommunications and Information Administration manages spectrum for federal users (47 U.S.C. §§ 303, 305). See GAO, *Commercial Spectrum: Plans and Actions to Meet Future Needs, Including Continued Use of Auctions,* GAO-12-118 *(Washington, D.C.: Nov. 23, 2011).*

[20] In the Matter of *Amendment of Parts 2 and 90 of the Commission's Rules to Allocate the 5.850-5.925 GHz Band to the Mobile Service for Dedicated Short Range Communications of Intelligent Transportation Services, Report and Order,* 14 FCC Rcd 18221 (1999).

[21] Radio frequencies are grouped into bands and are measured in units of Hertz, or cycles per second. The term megahertz (MHz) refers to millions of Hertz and gigahertz (GHz) to billions of Hertz. The Hertz unit of measurement is used to refer to both the quantity of spectrum (such as 75 MHz of spectrum) and the frequency bands (such as the 5.850 – 5.925 GHz band).

[22] *Amendment of the Commission's Rules Regarding Dedicated Short-Range Communication Services in the 5.850-5.925 GHz Band (5.9 GHz Band); Amendment of Parts 2 and 90 of the Commission's Rules to Allocate the 5.850-5.925 GHz Band to Mobile Service for Dedicated Short Range Communications of Intelligent Transportation Services; WT Docket No. 01-90, ET Docket No. 98-95, Report and Order*, 19 FCC Rcd 2458 (2004) (FCC 03-324).

[23] Traditional unlicensed equipment consists of low powered devices that operate in a limited geographic range, such as garage door openers and devices that offer wireless access to the Internet. They include Wi-Fi-enabled local area networks and fixed outdoor broadband transceivers used by wireless Internet service providers to connect devices to broadband networks.

[24] Middle Class Tax Relief and Job Creation Act of 2012, Pub. L. No. 112-96, § 6406, 126 Stat. 156, 231 (2012).

[25] In July 2013, the National Transportation Safety Board recommended that NHTSA develop minimum performance standards for connected vehicle technologies and require their installation on all newly manufactured highway vehicles.

[26] DOT defines light-duty vehicles as passenger cars, vans, minivans, sport utility vehicles or light pickup trucks with gross vehicle weight less than or equal to 10,000 lbs.

[27] The eight members of CAMP each provided eight vehicles that were fully integrated with V2V components for the Safety Pilot. The remainder of the vehicles involved were equipped with aftermarket and retrofit V2V devices. These devices were not fully integrated with the vehicle or not installed during vehicle production. In addition, 79 commercial vehicles and 88 transit vehicles were equipped with V2V devices and tested as part of the Safety Pilot.

[28] According to DOT, the department may release some information from its data analysis in relation to making this decision.

[29] For example, for potential collisions involving vehicles in a driver's blind spot, one automobile manufacturer's vehicles provided three short low-pitched beeps repeated three times, an orange light in the side view mirror, and a vibration on the side of the driver's seat in the direction of the potential collision.

[30] Public key infrastructure describes the hardware, software, people, policies, and procedures needed to create, manage, store, distribute, and revoke digital certificates. Public key infrastructure forms the foundation of the approach to V2V communications security by employing a "public key" to authenticate a sender or encrypt data being sent in a message, thereby producing trusted and secure messages. A public key infrastructure typically involves a certificate authority that issues and verifies digital certificates, which includes the public key, a directory that holds the certificates and a certificate management and distribution system.

[31] DOT has also been involved in efforts with the European Union and Japan to help set international standards for V2V technologies. For example, RITA has signed Implementing arrangements with both the European Union and Japan to cooperate on research programs and support global standards to ensure interoperability of systems. For more information see U.S. Department of Transportation *International Deployment of Cooperative Intelligent Transportation Systems – Bilateral Efforts of the European Commission and the United States Department of Transportation* FHWA-JPO-12-081 (Washington, D.C.: October 2012).

[32] Through "relative positioning," a vehicle's location is determined relative to other vehicles.

[33] V2V technologies normally transmit data 10 times per second.

[34] NHTSA views vehicle automation as progressing along various levels of automation from having only certain vehicle functions such as braking independently automated to the point at which the vehicle is fully automated and the driver provides no control over the vehicle during travel.

[35] KPMG and Center for Automotive Research, *Self-Driving Cars: The Next Revolution.*

[36] As noted previously, the Safety Pilot will end in February 2014. DOT plans to finalize its assessment and release its complete findings by fall 2014.

[37] The John A. Volpe National Transportation Systems Center, part of RITA, is a fee-for- service organization that performs work for DOT as well as other federal, state, local, and international agencies and entities.

[38] United States Department of Transportation, *Description of Light-Vehicle Pre-Crash Scenarios for Safety Applications Based on Vehicle-to-Vehicle Communications*, (Washington, D.C.: May 2013).

[39] According to DOT, based on early data on vehicle interactions in the Safety Pilot, there has been a sufficient level of vehicle interactions to analyze the performance capability of the applications, examine driver response to warnings, and evaluate unintended consequences.

[40] Insurance Institute for Highway Safety *Status Report,* Vol. 47, No. 5, (Arlington, VA.: July 3, 2012).

[41] After identifying a series of potential challenges facing the development and deployment of V2V technologies through preliminary interviews with DOT officials, two automobile industry groups, two automobile manufacturers, representatives of CAMP and the VIIC, and others, we asked experts identified by the National Academies of Sciences to discuss the extent to which each may pose a challenge. After interviewing 21 experts, in addition to automobile manufacturers and DOT officials, we counted the experts' ratings of potential challenges facing the development and deployment of V2V technologies and conducted a content analysis of their relevant responses. See appendix I for further details on our methodology and appendix III for a summary of the ratings provided by experts for each of the potential challenges discussed.

[42] U.S. Department of Transportation, *An Approach to Communications Security for a Communications Data Delivery System for V2V/V2I Safety: Technical Description and Identification of Policy and Institutional Issues*, FHWA-JPO-11-130 (Washington, D.C.: November 2011).

[43] U.S. Department of Transportation, *Enabling a Secure Environment for Vehicle-to- Vehicle (V2V) and Vehicle-to-Infrastructure (V2I) Transactions: April 2012 Public Workshop Proceedings*, FHWA-JPO-12-072 (Washington, D.C.: June 2012).

[44] Sec. 6406 of Act. Pub. L. No. 112-96, 126 Stat. 156, 231.

[45] In the Matter of *Revision of Part 15 of the Commission's Rules to Permit Unlicensed National Information Infrastructure (U-NII) Devices in the 5 GHz Band, Notice of Proposed Rulemaking, 28 FCC Rcd 1769 (2013).*

[46] 47 C.F.R. § 15.5.

[47] In its February 2013 Notice of Proposed Rulemaking, FCC stated that unlicensed devices typically operate at very low power over relatively short distances and often employ various techniques, such as dynamic spectrum access or listen-before-talk protocols, to reduce the interference risk to others as well as themselves. The primary operating condition for unlicensed devices is that the operator must accept whatever interference is received and must correct whatever interference is caused.

[48] See Department of Commerce, *Evaluation of the 5350-5470 MHz and 5850-5925 MHz Bands Pursuant to Section 6406(b) of the Middle Class Tax Relief and Job Creation Act of 2012* (Washington, D.C.: January 2013).

[49] FCC officials told us that they expect NTIA to finalize its recommendations for FCC by December 2014 and that, while FCC has generally stated that it would wait for the outcome of NTIA's studies before acting, FCC reserves the right, based on the record developed in the proceeding or changed circumstances, to act prior to that time.

[50] 78 Fed. Reg. 21320, April 10, 2013.

[51] As discussed earlier, warnings messages can be provided in a number of forms, including beeps or other audio warnings, bright lights, and seat vibrations.

[52] According to DOT officials, these principles address how to minimize the potential for driver distraction. They include a conceptual framework that will address, among other things, how to time the presentation of warning messages to drivers and how to prioritize multiple messages.

[53] For example, DOT plans to develop a driver-vehicle interface software-evaluation tool that will allow the automobile industry to evaluate the distraction potential of possible types of driver-vehicle interfaces. According to DOT officials, this tool will be finalized in 2015.

[54] Six clinics were conducted across the country between August 2011 and January 2012. The goals of the clinics were to promote V2V technologies, to assess their performance and reliability, and to obtain drivers' feedback. Over 90 percent of participants said that they would like to have V2V technology features in their vehicles and almost 80 percent reported having a high level of understanding of how these technologies work.

[55] We have previously reported on the challenges faced in protecting the privacy of consumer data collected by private companies, noting that entities have not consistently implemented recommended practices, such as notifying users about the collection of their data and placing limits on the retention of data, or consistently disclosed which third parties are given access to such information. See GAO, *Mobile Device Location Data: Additional Federal Actions Could Help Protect Consumer Privacy*, GAO-12-903 (Washington, D.C.: Sept. 11, 2012).

[56] However, such mobility applications may require the deployment of vehicle-to- infrastructure technologies.

[57] Consolidated Appropriations Act for 2005, Pub. L. No. 108-447, Div. H, §522, 118 Stat. 2809, 3268. The objective of a Privacy Impact Assessment is to determine if collected personal information data is necessary and relevant. These assessments are used to identify and address information privacy when planning, developing, implementing, and operating

individual agency information management systems and integrated information systems. They assess security and privacy risks associated with operating information systems that collect, access, use, or disseminate personal information.

[58] DOT Research and Innovative Technology Administration, *Enabling a Secure Environment for Vehicle-to-Vehicle (V2V) and Vehicle-to-Infrastructure (V2I) Transactions: April 2012 Public Workshop*, (Washington, D.C.: June 2012).

[59] As we have reported in the past, funding constraints have posed a significant challenge to the deployment of intelligent transportation systems technologies by state and local governments. See GAO, *Intelligent Transportation Systems: Improved DOT Collaboration and Communication Could Enhance the Use of Technology to Manage Congestion,* GAO-12-308 (Washington, D.C.: Mar. 19, 2012).

[60] While we interviewed Ford as part of group meetings with the Vehicle Infrastructure Integration Consortium and the CAMP VSC3 Consortium and interviewed a Ford official as an expert identified by the National Academies of Sciences, we did not otherwise individually interview Ford.

[61] These devices are wireless and include components such as a GPS chip to determine vehicle location and an antenna for data transmission at 5.9 GHz using DSRC. Vehicles with these devices transmit data on vehicle location and speed (called “basic safety messages”).

[62] Doing so would enable automobile manufacturers to earn higher government safety ratings for vehicles that support V2V safety applications. For more information on the New Car Assessment Program, please see www.safecar.gov.

In: Vehicle-to-Vehicle Technologies ... ISBN: 978-1-63117-045-4
Editor: Douglas Lacey © 2014 Nova Science Publishers, Inc.

Chapter 2

AN APPROACH TO COMMUNICATIONS SECURITY FOR A COMMUNICATIONS DATA DELIVERY SYSTEM FOR V2V/V2I SAFETY: TECHNICAL DESCRIPTION AND IDENTIFICATION OF POLICY AND INSTITUTIONAL ISSUES*

Anita Kim, Valerie Kniss, Gary Ritter and Suzanne M. Sloan

ABSTRACT

This report identifies the security approach associated with a communications data delivery system that supports vehicle-to-vehicle (V2V) and vehicle-to-infrastructure (V2I) communications. The report describes the risks associated with communications security and identifies approaches for addressing those risks. It also identifies and describes the policy and institutional issues that require focus in support of implementation and operations, as well as the balance needed among the

* This is an edited, reformatted and augmented version of Intelligent Transportation Systems (ITS), Joint Program Office (JPO), Research and Innovative Technology Administration, U.S. Department of Transportation publication, Report No. FHWA-JPO-11-130, dated November 2011.

priorities of security and safety with cost, privacy, enforcement, and other institutional issues.

The approach described in this report is a first step in identifying the technical and policy requirements that will form the basis for a prototype model that will be tested during the 2012-2013 Safety Pilot Model Deployment, located in Ann Arbor, Michigan. The prototype will be tested along with draft policies and procedures. Results of the test will inform the requirements, specifications, and guidelines for implementing an operational system.

EXECUTIVE SUMMARY

For 2010-2014, the primary focus of the United States Department of Transportation's (USDOT) Intelligent Transportation System (ITS) Program is a multimodal research initiative[1] focused on developing rapid and accurate wireless communication and data exchange among vehicles, roadside equipment (RSE), and passengers' personal communications devices. This innovative use of wireless communications offers an unprecedented opportunity to create an information-rich, connected vehicle environment that may transform surface transportation safety, mobility, and environmental performance.

The connected vehicle environment will use wireless, short-range communications to deliver data and create a dynamic data exchange between and among vehicle-to-vehicle (V2V), vehicle-to-infrastructure (V2I), and vehicle–to-mobile device (V2D). This exchange will support a variety of cooperative applications and systems.[2] Crash-avoidance safety applications are the highest priority and they establish the **minimum acceptable technical and policy requirements for security and trust** for a communications data delivery system. An additional important requirement is user acceptability which is based not only on the level of security but also on an appropriate balance between privacy, cost, and safety, among other important factors.

In considering technical and policy requirements as well as other requirements, it is worth noting that the envisioned communications data delivery system is establishing new ground as a system supporting safety- of-life using wireless communications. To date, wireless communications systems tend not to support safety-critical functions; nor do safety-critical systems tend to be based upon wireless systems. As such, the approach to security of the communications data delivery system draws from industry best practices but also establishes new practices to meet the specific needs and requirements of a connected vehicle environment.

Presented in this document is the proposed approach to communications security for a V2V/V2I safety communications data delivery system ("V2V/V2I communications system"). The objectives of this document are the following:

- Identify the **technical, policy, and institutional requirements for communications security** (based on USDOT, industry, and stakeholder needs);
- Describe the most probable **user and system risks**, their types and severity;
- Examine the varying levels of **security options available to address the risks**; and
- Examine the **policy and institutional issues associated with this approach and resulting impacts on safety, privacy, user acceptance, and cost**. It is expected that the most effective approach to communications security will leverage both technical and policy measures to optimally and effectively address security requirements.

ES.A. Approach to Communications Security—Requirements

The approach to communications security (herein, "the approach") was developed in partnership with industry[3] and with input from five teams of security experts whose expertise is based on experiences studying, designing, and implementing leading-edge security options for a variety of industries.[4] The approach draws upon globally-accepted best practices, but further configures these practices to meet the unique requirements necessary for a connected vehicle environment and, specifically, crash-avoidance safety applications. Four high- level, critical requirements form the foundation of the approach:

- ***Protection of Privacy*****:** The communications security system shall not allow for identification of a person through personally-identifiable information (PII) within messaging contents.
- ***Secure Communications*****:** All communications transmitted and received from a vehicle shall be secure. This includes both one-way and two-way communications. Messages will support delivery and

management of security credentials and will be encrypted to prevent eavesdropping and tampering over the communication channel.

- ***Trusted Communications:*** All communications exchanged between vehicles shall be trusted. Trust will be established through a user authentication process, which determines permissions and allowed actions with the system and other users.
- ***Scalability to Enable Nationwide Adoption:*** The security approach shall be scalable to support a population of over 250 million vehicles using the system.

In this approach, messages from an RSE are considered secure as the RSE will receive and use digital certificates to secure transmitted information. Other aspects of V2I security, however, were not included because they require further research. For example, hardware access and other elements of physical security for RSEs and aftermarket devices (ASDs) will require additional analysis, in particular as the system expands to include V2I mobility and environmental applications. At this point, it is expected that they will leverage similar concepts and elements from the security approach for V2V/V2I safety.

ES.B. Security Risks

There are two broad categories of risk associated with communications systems:

- **Attacks on the user/risks to safety and user acceptance:** Attacks on the user are aimed at directly impacting the safety of users and indirectly impacting system acceptance. With these types of attacks, attackers have two goals in mind: (1) to cause users to make bad driving decisions resulting in an accident, congestion, or reroute of a driver; and (2) to reduce users' faith in the system as messages become unreliable or unavailable.
- **Attacks on the communications system/risks to privacy:** Attacks on the communications system are attacks that lead to threats to privacy or cause drivers to bear extra administrative or legal burdens within the system. These types of attacks occur in two categories: (1) privacy attacks through tracking the location or driving route of a particular person; and (2) slander or framing attacks which involves

the false reporting of misbehavior from a vehicle, resulting in an otherwise valid driver being removed from the system.

Section II of this document will discuss in greater detail the expert evaluation of potential threats to the V2V/V2I communications system, an assessment of the probability of such attacks, and an analysis of the level of risk resulting from the threat. The overall conclusion of the expert analysis is that there are a low number of high risk attacks on the user and there appears to be minimal risk to safety in the event of a successful attack. This is, in part, due to the short-range nature of the communications and thus the geographic limitations to such attacks. Instead, analysis concludes that the greater risk is in reducing acceptance of the system to the point that users ignore it.

With regard to attacks on the system, the main risk appears to be related to privacy. To minimize this risk, the operational system is expected to use randomly assigned identifies that contain no identifying information. This configuration would make it necessary for an attacker to have some combination of inside knowledge of the system, physical access to the vehicle, or sizable investment resources to launch a successful attack on privacy. The complexity of such an attack is expected to deter attackers, reducing potential threats and ensuring that the risk to privacy is no greater than in today's world where attackers utilize more accessible methods such as cell phone tracking or physically following a vehicle.

ES.C. Approach to Communications Security—Configuration of Technical and Policy Options

Section III describes a set of technical and policy options designed to reduce risks specific to a V2V/V2I communications environment. The proposed approach combines three elements as the basis for a robust security solution:

1. Public Key Infrastructure (PKI)
2. Technical solutions (vehicle, hardware, and software)
3. Policy options

PKI is an umbrella term used to describe the hardware, software, people, policies, and procedures needed to create, manage, store, distribute, and revoke digital certificates.

PKI forms the foundation of the approach to V2V/V2I communications security by employing *public key cryptography* to authenticate a sender or encrypt data, thereby producing trusted and secure messages.

In addition to PKI, the approach integrates additional technical design elements (measures for providing security at both the vehicle and system level) and implementation of enforcement policies to deter attacks. The approach also employs techniques to detect misbehavior and remove the misbehaving entity.

In areas where security risks cannot be addressed through technical design alone, policy mechanisms such as governance, legal deterrence, and enforcement can serve a critical supporting role.

ES.D. Policy Issues Requiring Further Research

Section IV of this report summarizes the key policy and institutional issues that will need to be addressed in support of an operational system. The text additionally notes some of the inherent conflicts that may require decision makers and stakeholders to make choices or balance priorities.

The policy and institutional issues that are considered most significant (and thus will result in additional policy research) include:

- Analyses on how to most effectively design **organizational and operational entities** that will support security credential (certificate) management, legal deterrence, misbehavior detection, and revocation. Outstanding policy and institutional issues include questions on cost, whether to split the entities for enhanced privacy, personnel and equipment needs, and policies and procedures.
- Identification of the specific types of **legal deterrence and enforcement policies** that will act to prevent or mitigate misbehavior within the system. Outstanding policy questions include the determination of authority for enforcement.
- Development of a strategy for updating and **implementing the 2007 privacy principles**, including development of practicable options for putting the principles into use.
- Analysis on **implementation options** that compares different configurations using infrastructure and non-infrastructure options. Further, analysis on the types of **sustainable funding, financing, investment, and/or revenue sources** available with these

implementation options that address the needs for funding initial deployment as well as ongoing operations, and maintenance.

- The identification of the level and type of **governance and authorities** required for implementation of the organizational and operational models and with the communications data delivery system.

ES.E. An Evolutionary Path for Developing Communications Security

Development Process

Creating an approach represents the first step in the development process of designing and implementing a secure communications data delivery system for V2V/V2I Safety. The development process uses an incremental path that allows for review and decision making at critical junctures throughout research and development before a full system is deployed. This type of process encourages stakeholder input and provides opportunity for continuous refinement.

The steps of the development process are defined below.

- **Approach:** The approach represents the first step in identifying industry best practices and tailoring these practices to meet the requirements of a V2V/V2I environment for preventing, detecting, and mitigating security risks. The approach is the main focus of this document.
- **Design:** The design is the second step in the development process that will structure the requirements and technical elements into a representative prototype (due in Spring 2012). The design will assist in the targeted development of the policy, institutional, and organizational elements that support the operations of the system.
- **Model:** In this step, the technical prototype is combined with organizational and operational elements, resulting in a model system for testing and evaluation in a real world environment. The testing will occur during Safety Pilot Model Deployment (see side textbox for description; results from testing due in Summer 2013).
- **System:** The final step is to analyze the test results of the prototype model and to develop final technical and institutional requirements,

specifications, objective test procedures, and policies for an operational system (due in early 2014).

Further Research

The approach to communications security detailed in this document is configured to support successful initial implementation of crash-avoidance safety applications. From a technical perspective, it is possible for the approach to be scaled and expanded to support a wider range of safety, mobility, and environmental applications (e.g. notification of school zones, tolling, or fuel efficiency), although such an expansion is likely to require further technical, policy, and institutional research.[5] As noted previously, further research is also needed to address security for other elements that are external to the communications data delivery system. These elements include:

- **Security of devices:** mobile (personal) device security and aftermarket / retrofit device security;
- **Security of infrastructure:** security of communications infrastructure nodes (or roadside equipment), although it is not yet clear how much infrastructure will be needed; and
- **Organizational security:** security of the entity(ies) that manages the security credentials (referred to as a Certificate Authority, or CA) and associated policies for access to data and the system.

Although still conceptual, developing an approach at this time serves a number of important purposes. First, it allows all stakeholders involved to:

- Analyze the strengths, limitations, and vulnerabilities of the proposed approach
- Discuss the policy and institutional priorities and trade-offs; and
- Identify potential obstacles or challenges to implementation and operations.

Second, it allows stakeholders to provide feedback on how the approach supports their needs. Their inputs will be provided to the technical teams as the approach moves into a preliminary design and prototype. It also provides an opportunity to identify whether there are alternative approaches that may be more appropriate or effective.

Last, the development of the approach has highlighted the need to conduct policy research on key institutional elements that support implementation of an operational system. As noted above in section ES.D, research that is planned or underway includes:

- Development of options and definitions of proposed roles and responsibilities for **organizational and operational management entities**. This includes identification of enforcement techniques and governance needs that meet key requirements, such as privacy protection.
- Development of **viable financial models** to support initial implementation as well as ongoing operations and maintenance needs of the communications data delivery system.

ES.F. Organization of the White Paper

To present the proposed approach to communications security, this document is organized as follows:

- Section I: A description of the proposed approach.
- Section II: Analysis of the types of risks associated with communications security.
- Section III: An understanding of how the proposed approach is configured to address the identified risks. This section highlights the role of policy in combination with technical solutions.
- Section IV: Identification of the policy and institutional trade-offs. A preliminary trade-off analysis has resulted in an ability to generally consider security requirements vis-à-vis other priorities such as: cost, impact to safety, the level of institutional oversight and management, whether any new authorities are required for this approach, and enforcement needs.
- Section V: A description of the technical or policy gaps that still need to be resolved as the approach moves through the process of setting requirements, designing and developing a working prototype, testing, and implementing the system.

SECTION I: APPROACH TO COMMUNICATIONS SECURITY FOR A COMMUNICATIONS DATA DELIVERY SYSTEM FOR V2V/V2I SAFETY

For 2010-2014, the primary focus of the United States Department of Transportation's (USDOT) Intelligent Transportation System (ITS) Program is a multimodal research initiative[6] focused on developing rapid, and accurate wireless communication and data exchange among vehicles, roadside infrastructure, and passengers' personal communications devices. This innovative use of wireless communications offers an unprecedented opportunity to create an information-rich, connected environment for transportation that may transform surface transportation safety, mobility, and environmental performance.

The connected vehicle environment will use wireless, short-range communications to deliver data and create a dynamic data exchange from vehicle-to-vehicle (V2V), vehicle-to-infrastructure (V2I), and vehicle–to-mobile device (V2D). This exchange will support a variety of cooperative applications and systems.[7] Crash-avoidance safety applications are the highest priority and they establish the **minimum acceptable technical and policy requirements for security and trust** associated with the communications data delivery system. User acceptability is another priority and reflects the balance between the required level of security, privacy, cost, and safety. Acceptability, among other important factors, will form an important basis for adoption and use of the system and its safety applications.

In considering technical and policy requirements as well as other requirements, it is worth noting that the envisioned communications data delivery system is establishing new ground as a safety-critical system using wireless communications. To date, wireless communications systems tend not to support safety-critical functions; nor do safety-critical systems tend to be based upon wireless systems. As such, the approach to security of the communications data delivery system draws from industry best practices but also establishes new practices to meet the specific needs and requirements of a connected vehicle environment.

Presented in this document is the proposed approach to communications security for a V2V/V2I safety communications data delivery system. The objectives of this document are the following:

- Identify the technical, policy, and institutional requirements for providing communications security (based on USDOT, industry and stakeholder needs);
- Describe the most probable user and system risks, their types and severity;
- Examine the varying levels of security options available to address the risks; and
- Examine the policy and institutional issues associated with this approach and resulting impacts on safety, privacy, user acceptance, and cost. It is expected that the most effective approach to communications security will leverage both technical and policy measures to optimally and effectively address security requirements.

I.A. Objectives and Requirements

The approach to communications security (herein, "the approach") was developed in partnership with industry[8] and with input from five teams of security experts whose expertise is based on experiences studying, designing, and implementing leading-edge security options for a variety of industries.[9] The teams drew upon industry best practices, but adapted these practices to meet the unique requirements necessary for providing V2V/V2I safety applications.

In developing the approach, the analysis focused generally on two key questions –*What risks (security) are present and how can these risks be addressed? What technical and policy elements are required for a comprehensive communications security system?* In answering these questions, the team began to identify several issues influencing development of the approach, such as:

- Range of likely attacks and resulting risks to privacy, safety, and user acceptance of the system;
- Technical, policy, and legal options available to deter attacks;
- Processes for misbehavior detection, revocation mechanisms, and potential enforcement policies;
- Communication channel requirements for supporting secure data exchange between users and the system; and
- Options for establishing and maintaining trust through authentication, non-refutability, traceability, and auditing.

In partnership with stakeholders, the USDOT defined four, high-level but critical requirements that the proposed approach must meet in support of all of the V2V safety applications and a subset of the V2I safety messages. These requirements are necessary to ensure acceptance of the communications data delivery system. Safety of the user (and thus, security of the data and messages) was given the highest priority, with protection of privacy considered as a significant requirement for user adoption[10].

The critical requirements include:

- ***Protection of Privacy:*** The communications security system shall not allow for identification of a person through personally-identifiable information (PII) within messaging contents.
- ***Secure Communications:*** All communications transmitted and received from a vehicle shall be secure. This includes both one-way and two-way communications. Messages will support delivery and management of security credentials and will be encrypted to prevent eavesdropping and tampering over the communication channel.
- ***Trusted Communications:*** All communications exchanged between vehicles shall be trusted. Trust will be established through a user authentication process, which determines permissions and allowed actions with the system and other users.
- ***Scalability to Enable Nationwide Adoption:*** The security approach shall be scalable to support a population of over 250 million vehicles using the system

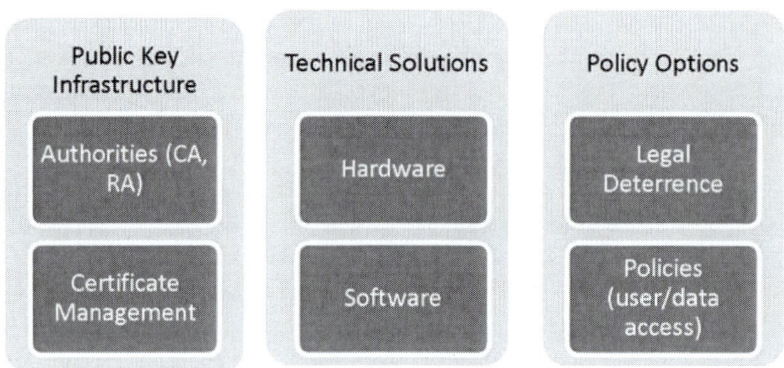

Figure 1. Key Elements of Proposed Security Approach.

In this approach, messages from an RSE are considered secure as the RSE will receive and use digital certificates to secure transmitted information. Other aspects of V2I security, however, were not included because they require further research. For example, hardware access and other elements of physical security for RSEs and aftermarket devices (ASDs) will require additional analysis, in particular as the system expands to include V2I mobility and environmental applications. At this point, it is expected that they will leverage similar concepts and elements from the security approach for V2V/V2I safety.

I.B. Configuration and Design

With objectives and requirements identified, the expert teams proposed three elements that provide the basis for the proposed security approach:

- Public Key Infrastructure (PKI) scheme
- Technical solutions (vehicle, hardware, and software security)
- Policy options

I.B.1. Public Key Infrastructure

PKI is an umbrella term used to describe the hardware, software, people, policies, and procedures needed to create, manage, store, distribute, and revoke digital certificates. As defined by experts, a PKI allows users of an unsecure public network to securely and privately exchange data through the use of a public and a private cryptographic key pair that is obtained and shared through a trusted authority. Although the components of a PKI are generally the same throughout many practices, a number of different approaches are employed to meet specific requirements.[11]

PKI assumes the use of *public key cryptography*, one of the most common methods for authenticating a message sender or encrypting a message. From an institutional perspective, a PKI typically consists of the following:

- A **certificate authority (CA)** that issues and verifies **digital certificates**. A certificate includes the public key or information about the public key;
- A **registration authority (RA)** that acts as the verifier for the certificate authority before a digital certificate is issued to a requestor;

- One or more **directories** where the certificates (with their public keys) are held; and
- A **certificate distribution and management system** which includes the communications system and its organizational and operational elements.

For the connected vehicle environment, PKI will be the basis for security and will be configured to provide a level of reliability, sensitivity, and redundancy needed for crash-avoidance safety applications. The primary function of the PKI is to allow users to exchange data through a trusted authority using authentication credentials (certificates). In addition to establishing trusted messages, the PKI will provide users with authorization credentials and facilitate certificate revocation in the event of misbehavior. While this framework for trusted communications meets the requirements for crash-avoidance safety, it is capable of being expanded to support mobility and environmental applications (e.g. tolling or parking reservations). Adapting this framework to fit non-safety applications will require further technical, policy, and institutional research.

Figure 2 illustrates the approach using basic PKI principles. It displays use of a certificate authority and a certificate distribution and management system for creating trusted messages between vehicles. Details of different PKI components and their application in the approach will be discussed in Section III.

- A one-way *trusted* communications between vehicles (V2V) supports the *broadcast of a Basic Safety Message (BSM)* once every tenth of a second, which will be received by any neighboring vehicles within range[12]. The BSM is used to exchange "vehicle state" data[13] for use in safety applications; thus, technical requirements include high latency, accuracy, reliability, and speed necessary for crash-avoidance safety applications to establish nearly immediate communications with other vehicles. The messages broadcast by vehicles are trusted, but not encrypted due to the time it would take to encrypt and unencrypt each message between vehicles, potentially hindering the immediate response times required for safety applications. This approach is the same for all safety messages that are transmitted to and from all devices, including safety messages from RSEs and ASDs.
- A two-way *trusted and secure communications between a vehicle and a certifying authority* may or may not require an infrastructure

> component. Two-way secure communications are encrypted and support the issuing of certificates and the certificate revocation list (CRL) which alerts users of misbehaving actors. These communications also support the reporting of misbehavior occurring within the system. At full deployment, it is expected that each vehicle will communicate with the CA at least once a day.

Under the approach, each new vehicle may have on-board equipment (OBE), which includes an on-board unit 808472219.g. computer module), display and DSRC radio. This equipment provides an interface to vehicular sensors, as well as a wireless communications interface to the CA. The on-board unit will store the vehicle's certificates which will be encrypted and grouped in batches. Certificates will only able to be decrypted, or unlocked, with keys that are issued from the CA.

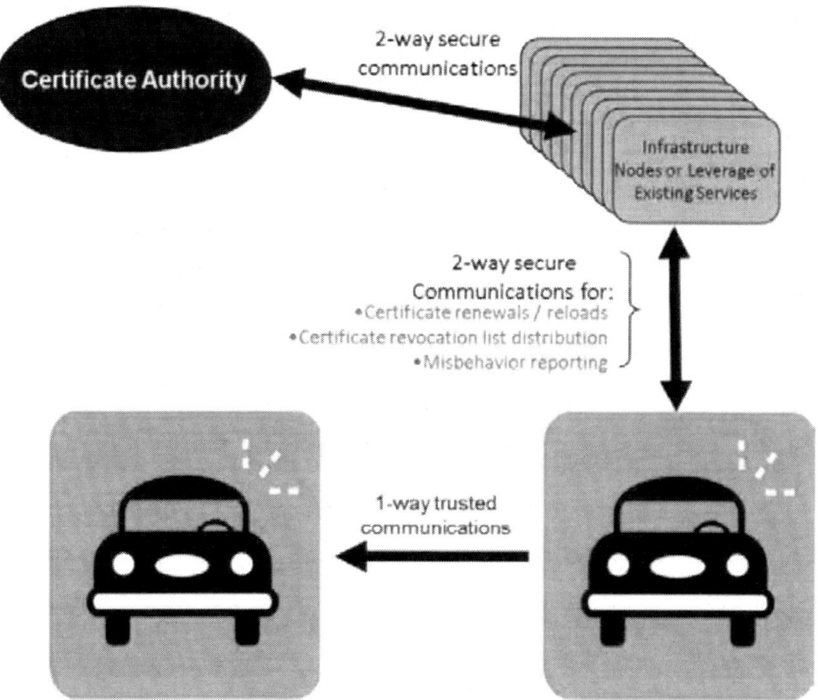

Figure 2. Proposed PKI Approach to Communications Security.

Options for the type of communications channel that will support the approach include dedicated short-range communications (DSRC), cellular or Wi-Fi, among others. Communications will go from the vehicle through an infrastructure node or use an already existing system to communicate with the CA. Ongoing policy research will analyze channel requirements as well as various business models capable of supporting channel use.

I.B.2. Technical Solutions

While PKI provides a framework for security, other technical elements can be incorporated at the local (vehicle) level to prevent misbehavior and detect misbehavior already occurring within the system. These technical solutions should work in combination with PKI and policies to make up the comprehensive security approach.

Security at the local level can be integrated into two different areas:

- Hardware: Includes such controls as a standard controller on the vehicle and tamper-proof encasements, among others; and
- Software: Includes functionality checks and misbehavior detection processes.

Hardware security supports restricting vehicle access and vehicle protection from tampering. Software security includes regular checks to identify misbehavior. Similar technical solutions will have to be applied for roadside equipment and aftermarket safety devices also. Further research on the specifics of RSE and ASD application is yet to be conducted.

I.B.3. Policy Options

Policies are an important component of the security approach. They provide non-technical solutions for deterring and addressing misbehavior. Such policies can include:

- Legal deterrence for physical tampering with vehicle's on-board equipment;
- Legal options for preventing misbehavior beyond revocation from the system;
- User access policies; and
- Policies that split the certificate management entity(ies) and ensure that accessing full information about any user will require the authority to gain access to multiple systems and sources.

I.B.4. Summary

Each of the key elements of the security approach (PKI, technical solutions, and policy options) is described in greater detail in later sections of this document. Similar elements must also be applied to roadside equipment and aftermarket devices in order to develop a comprehensive security approach. Further research on implementing a security approach for RSEs and ASDs is underway. In identifying the level of security needed for each element, an analysis of potential risks and threats was performed. These risks and threats are discussed in the following section.

SECTION II: SECURITY RISKS/THREATS

Identifying the type, probability, and potential impact of security risks or threats to the communications system was a critical first step in developing the proposed approach. By understanding the potential threat to the V2V/V2I communications system, an appropriate level of security could be designed into the approach.

The concept of *risk* can be interpreted in multiple ways. In this evaluation of risk, "system risk" is defined as the product of the probability of a successful attack taking place and its impact on the system. The probability, or likelihood, of a successful attack is correlated to two factors:

- **Overall cost to mount an attack:** Considers financial and time costs. It also accounts for the physical difficulties or complexity in mounting an attack as well as potential benefits.
- **Level of deterrence:** Refers to effectiveness of current laws and penalties if the attacker is caught in preventing or deterring attacks

Impact to the system requires quantifying the chance that an attack will lead to an accident, congestion, or route change, which depends on human response in a particular situation. Since determining human response is difficult, the impact to the system metric is converted into an estimated number of false messages that a driver will encounter. Overall, the aim of the approach is to reduce the number of false messages that a driver will encounter to decrease the level of safety risk. A review of threats to existing communications systems suggests that successful attacks would likely impact the following:

- User safety;
- Personal privacy;
- User acceptance of the system; and
- Communications operations.

Potential attacks on the communications system fall into two general categories: a) attacks on the user and b) attacks on the communications system. This section presents a summary of the expert evaluation of these types of attacks and their associated levels of risk.

II.A. Attacks on the User

Attacks on the user are aimed at directly impacting the safety of users and indirectly impacting system acceptance. With these types of attacks, attackers have two goals in mind:

- Cause users to make bad driving decisions resulting in an accident, congestion, or reroute of a driver; and
- Reduce users' faith in the system as messages become unreliable or unavailable.

While there could be additional motives for attackers to mount an attack on the user, these two represent the most common motives.

II.A.1. Attack Methods

To mount an attack on the user, an attacker must send messages via a radio in real-time to neighboring vehicles. These attacks are limited to the range of a radio, which is a 300 meter radius. There are three general methods an attacker can use to initiate an attack on the user:

- **Key Extraction:** Physically removing a vehicle's credentials (certificates and keys) from the on-board unit, and then using these credentials on other DSRC radios to create and distribute seemingly legitimate messages to neighboring vehicles.
- **Software Manipulation:** Installing malicious software on the vehicle's on-board unit to create messages containing arbitrary or altered information. The attacker can also manipulate existing software to extract certificates and keys.

- **Sensor Manipulation:** Interfering with the vehicle's sensor output to alter, inject, or suppress messages that originate from internal vehicle systems; or interfering with the sensor input that directly reports vehicle behavior or external circumstances.

The methods of attack noted above result in an attacker being able to create false messages which are then distributed to neighboring vehicles. They also require physical access and manipulation of the vehicle.

Another type of method for launching an attack on the user is **Denial of Service (DoS)** attacks. These attacks result in valid messages being suppressed or not received by the vehicle. These attacks do not require physical access to the vehicle, but do require access through a radio. They include:

- **Denial of Computation:** Sending large amounts of bogus messages[14] to a vehicle to cause the on- board unit to be overwhelmed with processing tasks, resulting in not receiving valid messages.
- **Denial of Communication:** Jamming the wireless band with a sufficiently powerful signal, denying vehicles the opportunity to transmit messages.

POLICY HIGHLIGHT

Key extraction poses the most significant risk to the system because the attacker is no longer restricted to attacks within the range of a single radio or vehicle. Key extraction allows an attacker to use multiple radios in multiple geographic locations.

Compromised keys can be used to create false messages, affecting user safety and system acceptance, but can also be used in privacy and framing attacks through attacks on the system Infrastructure.

There are existing laws for vehicle tampering which result in legal consequences. Policy research will explore the relevance of these laws in protecting a vehicle's on board equipment.

All vehicles have sensors that detect or measure a vehicle's movements such as acceleration, wheel movement, etc. The sensors transfer this collected information to the vehicle's internal computer in the form of wireless signals. Similarly, the sensors will transfer this information to the V2V applications as a basis for generating warning messages (V2V applications will not be used to override driver controls). Thus, a DoS attack has the ability to impair the V2V on-board equipment and applications, but not to impair the vehicle's internal computer (or Central Processing Unit, CPU).

II.A.3. Risk Level

Barriers exist on vehicles today that make launching a successful attack difficult. The chance of improper vehicle access occurring depends on the effectiveness of the vehicle's anti-theft solution, making some attacks no different from any other attack that can occur if a vehicle (or aftermarket device or piece of roadside equipment) is left unsecured. Thus, the threat of certain user attacks is similar to other more traditional threats (vehicle theft, equipment vandalism, etc.)

Also, an attacker must gain physical access to a vehicle for an extended period of time to launch a successful attack. The technological complexity of vehicle hardware and software provides a level of tamper-resistance. It is likely that only an attacker who is familiar with the vehicular architecture would be able to compromise a key, sensor, or GPS receiver.

It is possible that an attacker could extract keys or manipulate software or sensors on his own vehicle, however, as discussed in later sections, linkability of keys to owners will, in general, encourage attackers to focus on devices owned by other people as targets for key extraction. An attacker is unlikely to use his vehicle in an attack, as this may increase the chances of being caught. An attacker would most likely need to steal a vehicle or collude with a repair garage or rental car agency to gain access to a vehicle's on board unit.

Table 1 on the following page describes the methods for launching attacks on the user and the associated risk[15] to safety relative to an attacker's ability level. The risk levels take into account the theoretical implementation of technical elements recommended in the proposed approach. The risk assessment results in Table 1 do not factor in legal deterrence and enforcement methods that could provide an added layer of security to prevent against attacks on the user. Importantly, Table 1 evaluates risk based on the results of a comparison of the transportation environment with "no V2V/V2I system" versus a "perfect V2V/V2I system".

Table 1. Assessment of Safety Risk through Attacks on the User

Methods of Attack		Requires Physical Access	Level of Risk		
			A1	A2	A3
Key Extraction	Physically entering the vehicle and removing keys from the OBU	✓	Low	Medium	High
Software Manipulation	Use APIs[16] to extract keying material	✓	Low	Low	Medium
	Install software on device to create messages containing arbitrary information	✓	Low	Low	Medium
	Install software on device to alter information (i.e. system clock or sensor inputs)	✓	Low	Low	Medium
Sensor Manipulation	Interfere with input to CAN bus that directly report vehicle behavior (e.g. brakes)	✓	Low	Low	Medium
	Interfere with output from CAN bus to application processor	✓	Low	Medium	Medium
	Interfere with input from sensors that report external circumstances (e.g. GPS, lane marker detectors)		Low	Low	Medium
Denial of Service	Jamming the channel (denial of communication)		Low	Low	Low
	Send false messages that cause true messages to be ignored (denial of computation)		Low	Low	Low

The level of risk is dependent upon the ability of an attacker to carry about a successful attack. Attacker ability levels are as follows:

A1 = Clever Outsider – A talented engineer and/or cryptographer who does not possess any inside knowledge.

A2 = Knowledgeable Insider – An insider who possesses detailed knowledge about the system (security and non-security related) and has access to its specifications.

A3 = Funded Organizations – An organization that has access to substantial resources and furthermore possesses the capabilities of attacker A2.

II.B. Attacks on the Communications System

Attacks on the communications system are attacks that lead to threats to privacy or cause drivers to bear extra administrative or legal burdens within the system. These types of attacks occur in two categories:

- **Privacy attacks:** Tracking the location or driving route of a particular person
- **Slander/Framing attacks:** Falsely reporting misbehavior from a vehicle, resulting in an otherwise valid driver being removed from the system

II.B.1. Attack Methods

The attack methods for privacy and framing attacks require a level of technical manipulation of different components of the V2V/V2I system. Three major attack methods include: a) message linking, framing attacks, and Sybil attacks.

In **message linking**, an attacker sniffs V2V Basic Safety Messages, attempting to use information found within messages to identify a particular vehicle or a driver's whereabouts. The Basic Safety Message contains both static[17] and dynamic[18] identifiers, which can be used for message linking. These identifiers help build a dynamic map of the driving environment, which is used to predict vehicle movements, resulting in potential warnings to surrounding vehicles.

- *Syntactic Linking* – Uses static identifiers to establish multiple messages coming from the same vehicle
- *Semantic Linking* – Uses dynamic identifiers to "join the dots" to reconstruct a vehicle's trajectory using Vehicle Path Prediction methods

A **framing attack** is when an attacker makes a vehicle's on-board equipment appear to be malfunctioning by generating false messages to contradict the target vehicle's legitimate messages. This misbehavior is reported to the CA and could result in an innocent person's vehicle being revoked from the system. This requires an attacker to extract certificates and keys from another vehicle as described in Section II.A on page 19.

POLICY HIGHLIGHT

Privacy attacks use message linking, meaning that an attacker uses the contents of messages to determine a user's whereabouts. The approach proposes that messages do not contain any personally identifiable information, so that a person cannot be identified by an attacker simply by reading a message.

In developing the approach, it was assumed that **the approach should not compromise privacy any more than other systems that are in existence today**.

A **Sybil attack** is a sophisticated framing attack where an attacker simulates multiple vehicles by using multiple radios, allowing the attacker to send many messages by ultimately posing as several "ghost" vehicles at once. This creates a fictitious environment where it appears as if the targeted vehicle is misbehaving. An attacker needs to obtain valid certificates and keys from the vehicles to perform this attack, thus limiting the number of vehicles affected to the number of certificates obtained. Sybil attacks are a significant risk to privacy because they can be difficult to detect within the system since an attacker is using valid certificates and keys. Also, having multiple, simultaneously valid certificates on a vehicle increases the likelihood of a Sybil attack because an attacker can extract these certificates and use them at the same time to pretend to be multiple vehicles at once. The severity of a Sybil attack can be mitigated by restricting the number of concurrently valid certificates for a vehicle. A more in-depth discussion of the certificate management options is included in section III.A.2.3.

II.B.3. Risk Levels

Risk levels for privacy and framing attacks occurring in the V2V/V2I system relate to the appeal of existing methods for carrying out the same type of attack. The risk assessment in Table 2 compares the risk level of tracking a vehicle using the V2V/V2I system versus tracking a vehicle using existing or more traditional methods, such as tracking through cameras, cell phones, and tracking devices. Considering the cost and complexity of launching an attack through the V2V/V2I system, it appears more likely that an attacker would track a vehicle using existing methods, which can be performed at lesser cost and ease. Risk levels also vary between tracking a vehicle at a single location (e.g. person's workplace) and a larger area. An area risk could mean tracking a vehicle's route taken through an extensive area or to monitor overall driving behavior.

The risk assessment in Table 2 assumes certain security measures from the proposed approach are implemented, such as "anti-linking" measures. Anti-linking is a process that incorporates changing all static identifiers included within a message, including certificates, every five minutes, so that these identifiers cannot be linked together to track a person's whereabouts. No PII is contained in any static identifiers, eliminating the possibility that a person's identity can be lifted directly from Safety messages.

Incorporating anti-linking increases the complexity and investment required by an attacker to carry out a successful attack, resulting in lower risk to the system. Implementation of anti-linking significantly reduces the risk level when compared to a scenario where there is no anti-linking. The no anti-linking option is included only to illustrate the change in risk level associated with unmitigated message sniffing (meaning that no efforts are made to prevent linking of Safety messages).

In addition to risk from outside attackers, there is a risk of misbehavior from within the operators and overseers of the system. If the entities responsible for performing security operations were to collude or share information, personally identifiable information could be exposed or become vulnerable. Therefore, it is important that policies regarding data access and enforcement are in place to address individuals or entities wanting to misuse personal data.

II.C. Summary

Expert analysis suggests that the risk and impact to safety is generally low in user attacks and safety impact. The proposed approach makes it necessary for an attacker to have some combination of inside knowledge of the system, physical access to the vehicle, or sizable investment resources to launch a successful attack. Analysis concludes that the greater risk is in reducing acceptance of the system to the point that users deactivate it. With regard to attacks on the system, the main risk appears to be in terms of privacy. The risk to privacy is low as long as anti-linking mechanisms are implemented. An attacker would be more likely to utilize more convenient methods such as cell phone tracking or physically following a vehicle.

Table 2. Assessment of Privacy Risk through Attacks on Communications Infrastructure

	Single Location Risk			Area Risk		
	Preselect Person	**Check if preselect person is in data set**	**Derive person's identity directly from data set**	**Preselect Person**	**Check if preselect person is in data set**	**Derive person's identity directly from data set**
TRACKING VIA PROPOSED V2V SYSTEM (Anti-Linking Implemented)						
V2V message sniffing with no personal information used in identifiers & anti-linking	Low	Low	Low	Low	Low	Low
TRACKING VIA V2V SYSTEM (No Anti-Linking Implemented)						
Unmitigated V2V message sniffing	High	High	High	High	High	High
TRACKING VIA EXISTING METHODS						
Cell Phone tracking	Medium	Medium	Low	Medium	Medium	Low
Cameras	High	High	High	Medium	Medium	Medium
Physical tracking	Medium	Low	Low	Low	Low	Low
Tracking Device	High	Low	Low	High	Low	Low
RF Fingerprinting	High	High	Medium	High	High	Medium

Notes:

1) **Preselect person** – an attacker selects a particular person that they want to track
2) **Check if preselected person is in a collected data set** – an attacker could set up a tracking system, and then examine the data set collected from listening stations to determine if the preselected person appeared in the data
3) **Derive a person's identity from collected data -** an attacker may set up a tracking system, and then try to identify a person using only information obtained from within the collected data set

The next section describes how the technical and policy elements combine to form a communications security approach that meets V2V and V2I requirements. The description of the approach is further accompanied by a description of how the risks described in this section can be addressed through technical, policy, and a combination of both. The text in section III is annotated with options for system design and configuration.

SECTION III: ADDRESSING SECURITY RISKS—TECHNICAL AND POLICY OPTIONS

To achieve a robust communications security solution, the proposed approach combines both technical and policy options to address user and system risks. As described throughout this section, the approach first focuses on the deterrence of attacks through technical design and implementation of enforcement policies. The technical options include design measures for providing security at both the vehicle and system level. If these measures are circumvented, the approach then employs techniques to detect misbehavior and remove the misbehaving entity. As a last measure, where security risks cannot be addressed through technical design alone, policy mechanisms such as governance, legal deterrence, and enforcement can serve a critical supporting role.

III.A. Technical Solutions

The proposed approach addresses risks through the prevention and detection of misbehavior. Misbehavior is defined as either a malicious attack (as described in Section II) or a technical defect that occurs through mechanical malfunction. Options for preventing and detecting misbehavior exist at both the vehicle (local) level and through the communications system (global) level.

III.A.1. Mitigating Risk at the Local Level

III.A.1.1. Hardware Security

Hardware security is defined as the physical security features that protect the vehicle's on-board unit. Physical security features can protect against attacks that attempt to access the vehicle's on-board unit to extract keys or to manipulate a vehicle's software and sensors[19].

Security experts determined the level of hardware security required on the vehicle by analyzing the risk of successful attacks being launched on users and the system. Several options were compared in terms of cost and the time required for an attacker to compromise the on-board unit. Research indicated that if an attacker had prolonged access to a vehicle, most physical security mechanisms could be circumvented. Therefore, high-security hardware developed specifically for vehicles equipped for connected vehicle communications would not significantly increase the level of security. A standard controller[20] is proposed for providing hardware security with additional security from software operating on the hardware, rather than through dedicated hardware alone (see section III.A.1.2 on the following page). In addition to the standard controller, it is important to add a level of physical security on the exterior of the on-board unit that makes it difficult for an attacker to extract keys from an on-board unit undetected in a small number of hours. This could be as simple as providing a tamper-evident seal to show unauthorized physical access, similar to gas and electric meters.

Because none of the commercially-available and cost-feasible hardware options can completely prevent key extraction and other physical security risks, it proposed that a system of legal deterrence against physical tampering is used as a countermeasure. This could be similar to deterrence implemented for odometer fraud or tampering with brake lines. Further details on legal deterrence are presented in the next section.

POLICY OPTION

None of the currently viable commercial hardware options can completely prevent key extraction and other physical security risks once an attacker has gained access to the vehicle. As a countermeasure, legal deterrence and enforcement are needed to compliment the technical measures to protect against equipment tampering.

Since this approach may be extended to incorporate other V2I safety, mobility, and environmental applications, it is recommended that safety and non-safety related applications run on separate hardware platforms. Both types of applications would use different controllers but share vehicle infrastructure, such as the radio and vehicle computer[21]. Even with this segregation, attacks on non-safety applications could result in Denial of Service (DoS) attacks for the safety applications if there is shared infrastructure supporting the communications. Additionally, there may be cost implications. Nevertheless, the separation of the processing platform is expected to prevent basic attacks on non-safety applications from compromising the safety applications.

POLICY DISCUSSION

As noted previously, key extraction from the on-board unit poses the most significant risk, with denial of service and denial of communications (jamming) following.

Does the serious nature of these attacks require legal deterrence that breaks with the principles of privacy protections (i.e., implementing the ability to identify misbehaving actors)?

III.A.1.2. Software Security and Misbehavior Detection

The combination of hardware and software implemented at the vehicle level is referred to as the hardware architecture. While hardware security aims to prevent attacks, the addition of software functionality checks and misbehavior detection processes *allows for the capability to detect misbehavior that has successfully circumvented physical security measures and entered the system.* (Misbehavior is defined as either a malicious attack, such as false messages that have been successfully injected into the system or a technical defect that occurs through mechanical malfunction[22].) Implementing this capability is especially important for detecting attacks that cannot be mitigated through physical security, such as Denial of Service attacks which are launched remotely.

If an attacker successfully circumvents the physical security measures on a vehicle, several software functionality checks running on the hardware will detect equipment tampering. Software checks at the vehicle level include:

- **Secure flashing:** is a process that protects the software running on an Electronic Control Unit (ECU)[23], by ensuring that only authorized software can run on the ECU. In this case, the ECU is the on-board unit. Secure flashing combats software manipulation attacks by detecting software that was installed by an attacker designed to create false messages, alter sensor input information or extract keying material. This also ensures that the core functionality of the DSRC radio cannot be compromised.

- **Component identification:** is a process used by the ECUs in a vehicle to identify each other. ECUs establish a "home environment" and if an ECU is taken from one vehicle and put into another vehicle, it stops functioning. It does not prevent attacks but adds another layer of difficulty the attacker has to overcome. By allowing the DSRC radio to participate in the vehicle's component identification scheme, this would prevent the DSRC radio from functioning outside the intended vehicle, mitigating attacks where the radio is stolen or removed from a junked vehicle and directly installed into another vehicle.
- **Plausibility checks:** detect bogus sensor data generated from either manipulated or faulty sensors on both the sender and receiver vehicles. These checks run continuously on all messages between vehicles. If the vehicle can detect that the message received is bogus or spurious[24], then it will not relay this message to other vehicles. It is expected that almost all technical defects would be able to be detected by the on-board diagnostics of the sending vehicle and defective messages would not be sent out. One way that a vehicle checks plausibility is by building a physical model of its surroundings and checking it for consistency – known as location checks. Consistency checks can also be internal to ensure that the current state of the on-board unit is consistent with the last known state. Denial of Communication attacks cannot be prevented through physical security but can be detected through plausibility checks.

III.A.1.3. Misbehavior Detection at the Local Level

Random message checking is a process that collects messages from vehicles to be sent to the certificate authority for further checking. First, local processing checks the validity of each message through plausibility checks at the vehicle level. Local processing is performed without active interaction with other vehicles or with the certifying authority. After verifying incoming messages, the vehicle then collects a random number of messages (referred to as reports) to be sent to the CA for further checking, known as global processing. Reports could include a record of an event as well as random non-suspicious messages. This builds a global view for detecting misbehavior. This is an important tool for detecting Sybil attacks since the CA is collecting reports from many vehicles at once and can detect that certificates apparently used by different vehicles were all in fact extracted from the same OBU (using authority traceability). This process of misbehavior detection allows for certificate revocation of bad actors, due to both technical defects and malicious attacks. More discussion of global processing and authority traceability can be found in section III.A.2.5 on page 34.

POLICY DISCUSSION

To detect misbehavior and appropriately revoke a user from the system, the certifying authority must use "authority traceability" to identify the misbehaving entity. A decision must be made regarding how far back in time messages can be linked to a vehicle by the authority.

Depending upon how the authority decides to enforce misbehavior, either through simple revocation from the system or other legal options, this could reveal a person's identity and impact privacy.

It is expected that some false messages will manage to circumvent the software and misbehavior checks. Using the proposed configuration, estimates are that a user will encounter a false message once every four days. Although there has not been extensive research conducted on how drivers will react to a false message, initial research suggests that this type of error does not pose a significant risk to safety. The driver is ultimately in control of the vehicle and can ascertain (in most situations) whether it makes sense to react to a message. The greater risk of false messaging is that it could ultimately reduce driver acceptance of the system. Drivers could choose to ignore accurate messages because of the assumption that some messages are providing false information. This behavior may pose a safety problem in that a driver may ignore an accurate message that could have prevented a collision.

POLICY DISCUSSION

Is the **encounter with a false message once every four days** an acceptable level for public acceptance of the system if the level of risk to safety is low?

Suppressed messaging is the result of Denial of Service (DoS) attacks and refers to instances when a warning should have been issued in a scenario (e.g. crash avoidance application), but did not occur. The near-term consequences of suppressed messaging can cause an outage of collision avoidance warnings, which returns drivers to the level of awareness that existed when there was no system. However, these risks may increase over time as users become accustomed to relying upon generated warnings and thus are likely to place less weight on his/her own evaluation of an imminent collision within the surroundings. To counteract this risk, if an attack is detected the approach suggests that the receiving unit issue a "Service Temporarily Unavailable" message to protect a driver from getting confused between the absence of a message due to no danger, and the absence of a message due to channel congestion resulting from a DoS attack.

III.A.2. Mitigating Risk at the Global Level

Protecting privacy at the Global level includes options for implementing "anonymous, randomly-assigned identifiers with each message" and further ensuring that messages are "unlinkable". Anonymous identifiers mean that a user cannot be identified through content included in the messages. A user is said to have unlinkability when it cannot be determined that two messages belong to the same user.

III.A.2.1. Public Key Infrastructure

A Public Key Infrastructure (PKI) to establish trust within the communications network. Its purpose is twofold: to provide authorization credentials to users for participation in the network and to revoke those credentials if an administrating authority decides to do so. The PKI design incorporates the use of authorization credentials to support trusted communications between users. Authorization credentials in a PKI are known as certificates, which represent information cryptographically bound together and certified by an administrating authority called a certificate authority (CA). Although this approach is presented in terms of vehicles communicating with the certificate authority, it is proposed to be applicable to other mobile users[25] during full deployment.

III.A.2.2. Certificate Authority

An important component of the approach to security is a Certificate Authority. A CA is an authorizing organization that issues and revokes certificates, and maintains and distributes certificate status information. It is generally a trusted party that is central to the security services of the system. A CA can detect misbehavior occurring within the system as a result of technical malfunction or intentional malicious attack. In either case, the CA can revoke the certificates associated with this misbehavior and inform other users within the connected vehicle environment that these revoked certificates are no longer trusted.

Communications between vehicles and the CA must be trusted and secure. "Secure two-way" in this context means establishing a communications session with trusted credentials at each end that use encryption of messages sent over wireless communication and backhaul technologies to prevent eavesdropping and tampering. In the proposed communications exchange (see Figure 2 on Page 15Error! Bookmark not defined.), the CA issues certificates, private keys, and information about revoked certificates to the vehicle through a communications infrastructure node or existing network, allowing the vehicle to obtain new certificates and keys, protect itself against bad actors, and ensure that communications with other vehicles are trusted.

The expert teams evaluated several communication scenarios that take into account the amount of bandwidth and access time required for communication with the CA. Greater available bandwidth and lower time-to- access provides for a more effective the security mechanism. However, cost and privacy considerations prohibit continuous connection of vehicles to the CA.

POLICY DISCUSSION

A CA insider or the CA itself could be untrustworthy or act maliciously. This risk is addressed by dividing responsibilities and information across multiple CAs or between a CA and a Registry Authority (RA). No one authority would hold enough information to link a certificate or message to a vehicle identifier or actual driver identity.

The segregation of information across multiples entities carries cost implications. Analysis to date suggests that the institutional costs are likely to be small by comparison to the gains in privacy protection and public acceptance. The policy research effort to develop organizational and operational models (options) will provide greater detail on costs and effectiveness associated with splitting the CA into multiple entities (See Appendix A).

Proposed Organizational Structure for Certificate Authority

There is a privacy risk associated with the CA, stemming from the fear that the CA itself could be untrustworthy or act maliciously. The proposed approach recommends incorporating a "split CA", which divides the certificate authority's capability and functionality by both technical and organizational means in order to enhance privacy. Splitting the CA creates a situation where one administrating authority does not hold enough personally identifiable information to compromise a person's privacy or vehicle's identity.

As a hypothetical example, the overall CA structure can be divided into a Request and Registration Authority (RA) and a Certificate Authority (CA). The RA verifies requests for keys and certificates and authorizes assignment of keys and certificates. It determines whether a certificate request should be granted, but will not be involved in creating and assigning the keys and certificates to vehicles or other mobile users. The CA creates, stores, and assigns keys and certificates to vehicles authorized by the RA. The CA also is responsible for credentials management, CRL management, and receiving misbehavior detection reports. The CA recognizes invalid certificates from the misbehavior detection agent, adds the vehicle revocation identifier on the CRL, signs the CRL, and performs all functions of key management.

Splitting functions between entities could have cost implications, although the expert teams indicated that this cost would be minimal and without impact to the communication channel requirements. There is a research effort being launched to develop organizational and operational models (options) for the certificate authority that will provide greater detail on costs and effectiveness associated with splitting the CA.

Policy Discussion

User safety and misbehavior detection may require the CA to maintain short-term linkability with the vehicle. Further stakeholder discussion is needed to determine the appropriate balance among these safety requirements with the importance of privacy.

III.A.2.3. Certificates

Certificates are necessary for users to verify that other users are authorized to use the network and can therefore be trusted. To protect user privacy, certificates do not contain identifying information, but instead use random pseudonyms as temporary identifiers in messaging. However, even with this measure in place, using a single certificate over a large time-span would allow long-term tracking of a vehicle. Therefore, the approach proposes that each vehicle is loaded with a set of certificates that are changed regularly. Privacy is enhanced with more frequent changes but this also increase local storage and communication requirements.

Proposed Certificate Management Approach

Several options for the certificate were identified with the intent to promote maximum safety benefits, but to also protect user privacy by supporting anonymity and unlinkability.

Each vehicle is equipped with two different types of certificates:

- Short-Lived Certificates (SLCs) – These certificates are used during normal mode of operation and are changed every five-minutes.
- Long-Term Certificates (LTCs) – These certificates are used during fail-safe mode during low- likelihood events when SLCs may not be available.

The following is the proposed certificate protocol:

- Each SLC is only valid for a single, predetermined, time-restricted five-minute period. Each certificate will have a fixed start and end time and is valid only during the specified period, regardless of whether it is actually used during that time or not.
- All vehicles share the same time restricted intervals so an attacker is not able to distinguish between any two vehicles based on its start and end times. As an example, the CA issues all certificates so that the first certificate is valid on January 1, 00:00:00 – 00:05:00, 00:04:30 – 00:09:30, and so on. Fixed times also increase the protection against Sybil attacks by limiting the number of certificates available at any given time.
- SLCs will overlap with preceding and following certificates for 30 seconds to ensure that the vehicle always has access to a valid certificate. It is believed that a 30 second overlap provides enough time to avoid any outages due to failed certificate changes, as well as ensuring communications during critical circumstances, such as a certificate change delay during a pre-crash safety communication.
- All static identifiers[26] contained in the Basic Safety Message, including the certificate's temporary identifier, will be changed simultaneously every five minutes. Changing certificates and other identifiers often is important to protecting privacy because it makes tracking a vehicle via information contained in unencrypted Basic Safety Message (BSM) more difficult.
- A certificate's temporary identifier field within the Basic Safety Message (BSM) will be randomized. Other static identifiers, such as

vehicle length and width, will not be randomized, but instead will be specified loosely enough that a particular vehicle cannot be identified by these types of static fields. [27]

POLICY DISCUSSION

The approach includes changing certificates every five minutes. This time period was selected based on research that estimated a **7% chance of an attacker compromising privacy**. A validity period of two minutes essentially avoids privacy compromise, but presents a higher cost associated with certificate storage and communication with the CA.

The five-minute validity recommendation was established based upon discussion among the security experts when evaluating the trade-off between privacy, safety, and security. A 10-minute validity gives the attacker a likelihood of almost 28% to compromise privacy. A five-minute validity gives the attacker a likelihood of around 7% to compromise privacy. A two- minute validity essentially avoids privacy compromise. However, a two- minute validity could begin to have a more significant negative impact on safety since an identity change temporarily weakens a vehicle's ability to track surrounding vehicles. Furthermore, switching certificates below the five-minute threshold could compel an attacker to use less expensive and technologically advanced methods of tracking, such as radio frequency fingerprinting. Therefore, it was deemed that the five-minute certificate validity period to be a feasible approach. This is a technical recommendation but does require agreement from privacy advocates that this is an appropriate approach.

The proposed approach also addressed instances when short-lived certificates may not be available due to time restricted certificates expiring. An example is when a user is on vacation for an extended period of time and has not been in communication with the CA and is therefore unable to unlock the next batch of short-lived certificates. Vehicles will go into fail-safe mode and use LTCs in these cases. The availability of long term certificates assures uninterrupted participation in the system. Thus, upon return to driving, the vehicle's ability to use safety applications will not be compromised due to an expired certificate and safety applications will function until the vehicle unlocks new SLCs. This has a temporary impact on privacy since a vehicle could be tracked for longer periods through the use of the LTC. It has not yet been determined whether one or more long-term fail-safe certificates per vehicle will be required; or the validity time period for these certificates. The availability of long term certificates also increases the number of available certificates to an attacker, thus increasing the likelihood of a successful Sybil attack and risk to privacy.

III.A.2.4. Certificate Management and Distribution

The manner in which the CA issues and revokes certificates is defined as the Certificate Management scheme. The proposed approach uses Linked Certificates – meaning that certificates will be cryptographically linked together by an encrypted identifier. The certificates are stored on the vehicle and are encrypted until unlocked by a decryption key. The vehicle's on-board unit must request this key from the CA. Each key request would unlock a different group of certificates. This reduces the possibility that a bad actor can gain access to a large number of certificates at one time, thus reducing the likelihood of a Sybil attack.

Since the use of time-restricted certificates introduces the need for a greater number of certificates per vehicle, all certificates will be encrypted and loaded onto the vehicle at longer-term intervals, for example, annually. The certificates are stored in batches, or bundles, and cannot be unencrypted until a key is received from the CA. At full deployment, vehicles will request keys from the CA daily to unlock certificates. Vehicles will receive certificates, keys and CRLs through infrastructure nodes, most likely deployed as roadside units (RSEs).

Revocation information will be distributed in the form of a certificate revocation list (CRL) by the CA. Each revoked vehicle requires only a single CRL entry containing the small revocation key since all certificates are linked, as compared to putting all individual revoked certificates on a CRL. This decreases the size of the CRL significantly, reducing the bandwidth required for CRL communications. The CA can then deny requests for new certificates or decryption keys based on this list. This recommendation allows vehicles to be informed sooner, but requires more communication (in terms of bandwidth) between vehicles and the CA for aspects of certificate management other than CRL distribution.

III.A.2.5. Misbehavior Detection at the Global Level

Options for misbehavior detection exist at the local level and at the global level. As discussed earlier, some attacks can be detected at the vehicle (local) level through software functionality checks and other misbehavior detection processes. While local level processes result in a vehicle not accepting false messages, it does not address the removal of the misbehaving actor from the system. This is a global level misbehavior detection process. Global processing provides a method for gathering a system-wide view of misbehavior to detect attacks that may not be geographically limited. Global processing incorporates random message checking to collect a certain number of local messages generated by the vehicle (referred to as reports) to be sent to the CA for further checking. Misbehavior reports are collected from many vehicles and compiled to determine whether a reported vehicle is misbehaving through analysis of message content, sensor data, onboard dynamic state map, or physical laws. Since the CA is collecting reports from many vehicles at once, the CA can more easily detect Sybil attacks (as compared to local processing detection) by detecting that certificates apparently used by different vehicles were all in fact extracted from the same OBU.

POLICY DISCUSSION

Frequent vehicle reports to identify misbehavior increases security by allowing the authority to quickly discover bad actors within the system. The trade-off may be increased risk to privacy, as linkability is required to take action on a misbehaving vehicle.

Also, collection of more frequent reports is a trade-off between increased security and more CA communication which may increase congestion on the channel as well as increase CA administration costs.

Options surrounding what information will be included in the vehicle reports, how often reports are collected, and what constitutes misbehavior will be included in the analysis for organizational and operational models.

In order for the CA to revoke misbehaving vehicles from the system, linkability is needed to allow for messages sent by a vehicle to be linked to the vehicle's cryptographic identity. This is known as *authority traceability*. The authority traceability property makes the reporting and the reported vehicles accountable to the CA. The actual decision of whether the reported vehicle is malicious or a victim of a slander attack would be made by the CA after examining all of the available contextual information. Authority traceability may increase the risk of an insider tracing a particular vehicle, but this risk can be mitigated through the split CA functions.

III.A.2.6. Communication Network – Bandwidth and Access Time Requirements

With an understanding of the risks and mitigation features described throughout section III, the remaining analysis focused on the frequency of a vehicle's communication with a CA to update certificates, the revocation list and to report misbehavior. While frequent communications increases security, there is a direct trade-off between access time for vehicle and CA communications and bandwidth usage.

The analysis did not assume a specific communications media when developing the requirements for access time and bandwidth. Instead, it looked at the amount of bandwidth required for security credentials management based on the potential vehicle population (assumed at 250 million vehicles as the basis of a nationwide deployment).

Certificates linked to misbehavior should be revoked quickly to reduce the window of opportunity for misbehavior to damage the network. It is assumed that there will be fewer instances of attacks with less deployment. Three levels of deployment were considered in establishing the Time-To-Access (TTA)[28] and the time it takes to revoke an attacker—10 percent deployment, 50 percent deployment, and 100 percent deployment. The results are in the table below:

Level of Deployment	TTA	Time between the start of an attack and the revocation of an attacker
10%	Every 10 days	25 days
50%	Every 2 days	5 days
100%	Once every day	2.5 day

POLICY RESEARCH

Research is needed to establish criteria for an effective process for end-of-life for on-board equipment. Long- term certificates (LTCs) could be valid for a vehicle's lifetime. Further policy research is needed on the implications of "junking" a vehicle or transferring ownership. If the LTCs are not revoked at the end of the vehicle's life, this opens the possibility of attackers stealing certificates from junked vehicles. Such issues will be explored further in policy research that seeks to identify legal enforcement options.

With this analysis, bandwidth requirements and the ideal options for how and where a vehicle communicates with a CA (for instance, on highways, local arterials, at gas stations, etc.) can be determined.

SECTION IV. PRELIMINARY POLICY ANALYSIS

As noted in the side textboxes throughout Sections I-III, technical design options may rely upon policy solutions to provide comprehensive security. This section identifies the policy and institutional elements and describes the type of policy research being conducted in support of the approach. The text additionally notes some of the inherent conflicts that may require decision makers and stakeholders to balance priorities.

The policy and institutional issues that are considered most significant (and thus will result in additional policy research) include:

- Analysis on how to most effectively design organizational and operational entities that will support security credential (certificate) management, legal deterrence, misbehavior detection, and revocation. Policy and institutional issues include questions on cost, whether to split the entities for enhanced privacy, personnel and equipment needs, and policies and procedures.
- Identification of the specific types of legal deterrence and policies that will act to prevent or mitigate misbehavior within the system. Policy questions include the determination of authority for enforcement.
- Development of a strategy for updating and implementing the 2007 privacy principles, including development of practicable options for putting the principles into use.
- Analysis on implementation options that compares different configurations using infrastructure and non-infrastructure options. Further, analysis on the types of sustainable funding, financing, investment, and/or revenue sources available with these implementation options that address the needs for funding initial deployment as well as ongoing operations, and maintenance.
- The identification of the level and type of governance and authorities required for implementation of the organizational and operational models and with the communications data delivery system.

IV.A. Organizational and Operational Models for a Certificate Management Entity

A technical solution for greater privacy protection involves the use of a split CA. Splitting the CA creates a situation where one administrating authority does not hold enough personally identifiable information to compromise a person's privacy or vehicle's identity. By splitting the CA functions, multiple organizations must intentionally collude and act together to negatively impact personal privacy.

In addition to assessing the options for how the organizational and operational entities might be configured, further policy research is needed to address:

- The appropriate level of institutional resources that are needed and whether existing institutional arrangements can be leveraged. Included is an analysis of the public/private sector roles and determination of whether the entities must be wholly public, can be wholly private, or require some level or partnership;
- The costs associated with security credentials management as well as organizational and operational costs; and
- The governance authority needed to provide legal enforcement.

In determining how to structure a CA for V2V/V2I, policy research is underway and will result in the development of options for:

- Configuration of a certificate management entity(ies) and how they might split functions across multiple authorities. The options will be accompanied by proposed policies for enforcement and for overseeing the separation, and definitions for clear roles and responsibilities within the organizational structure(s).
- Policies and procedures regarding misbehavior—whether to simply remove the misbehaving actor or to further link the actor to a personal identity in order to legally address the misbehavior and/or revoke the privileges of certain users or vehicles. While linking the action to a person provides a more complete security approach, it violates the principle of anonymity. Policy decisions must still be made on how this policy option should be implemented, especially in relation to the overall security context.

- Governance options and analysis of public and private roles and/or partnerships.

Further detail on this policy research is provided in Appendix A.1.

IV.B. Legal Deterrence and Enforcement

The proposed security approach must assume some level of legal deterrence and enforcement to address potential bad actors. As shown in other industries, strict legal consequences paired with effective enforcement methods can deter bad actors and strengthen the overall security approach.

For V2V/V2I communications, legal deterrence is especially important for limiting improper physical access to vehicles. Limiting physical access is essential to a vehicle's hardware and software security. Similar to how illegal access and tampering to personal vehicles are currently addressed through legal enforcement, a comparable policy technique could be used for vehicles equipped with connected vehicle safety technologies and applications.

Legal deterrence and policies for enforcement are also critical to addressing misbehavior. The current design proposes that misbehavior can be detected, resulting in the bad actor being removed from the system (through revocation of certificates). An additional option would be to identify the misbehaving party(ies) in order to take further legal action. This would require cooperation between the split CAs for identification, but would also violate the privacy principles as they stand today. Included with the research on certificate management entities (Appendix A.1) is research to identify which actions would be considered illegal and the types of enforcement techniques available within the limitations of the system design and the privacy principles. A separate legal analysis will be performed to classify the legal and illegal actions and to determine how best to address them from a user perspective while considering concerns about privacy and crime.

IV.C. Privacy

As defined today, the security approach provides three important conclusions regarding privacy when implemented in support of vehicle-to-vehicle communications for crash-avoidance safety applications. They include:

1. No personally-identifiable information (PII) is available in the system and thus, an attack on the operational system cannot violate privacy
2. Short duration of tracking is possible but the attack cannot obtain vehicle identity through the system and the security approach makes it *very difficult for this to occur (requires complex equipment which means a sizable investment)*
3. Vehicle identification is available only through integration of information from each certificate authority, *a risk that will be addressed through policies on access to systems, hiring/ employment checks, and proper enforcement in place*

The above conclusions will need to be revisited from the perspective of mobility, environment, and convenience applications that may require a user to identify a device (i.e., when requesting routing information) or provide other information (such as financial information for parking). Other remaining issues that still require analysis are:

- Given the requirement of anonymity by design, what are the implications for cost, communications load, and deployment?
- Given the requirement of anonymity by design and the privacy principles, would it ever be appropriate to identify an individual as a bad actor? Is this trade-off acceptable to the public in terms of using and trusting the system?
- How effective are existing policies on system access, hiring/employment, and improper use of data, based on lessons learned from other industries? (will be addressed through research described in Appendix A.1)
- Is this approach to privacy acceptable to privacy advocates, safety advocates, and the public?

IV.D. System Implementation: Costs and Sustainable Funding/Financing/Investments and Levels of Security

Identifying the ability to finance system implementation and provide sustainable funding for operations and maintenance is another critical policy issue under analysis. The level of system installation and ongoing costs may vary significantly based upon the type of deployment or implementation model used. For instance, a system model that requires new infrastructure is likely to cost much more at implementation than one that leverages existing infrastructure or other equipment (for instance, cellular or Wi-Fi networks).

There are three important elements to considering costs—costs for developing the security approach, costs associated with implementation, and the ongoing cost of operations and maintenance (importantly, in terms of operations, most communications security systems are not designed and implemented on the scale envisioned for the V2V/V2I system). An important element to consider when identifying options for financing, partnerships or investments is whether options exist to create a revenue stream that would help sustain ongoing costs. Thus, further research into viable models is needed. Appendix A.2 describes the research efforts associated with these requirements.

IV.E. Governance

Technical solutions play an important role in mitigating risk, deterring misbehavior, and protecting privacy. Policy and institutional solutions result in institutional processes for organizations, addressing user access, rules of operations, enforcement and other important elements of a robust operational system. Policy research is important for determining whether the implementation and ongoing operations of a secure communications data delivery system is viable.

Eventually a set of decisions will be needed to determine who will take on the roles and responsibilities for implementation, operations, oversight, maintenance, and conflict resolution. This question falls under the issue of governance. Other governance questions include, but are not limited to:

- Can governance (as well as the institutional solutions for CA and security) be wholly private, or must it be wholly public, or some mix, and what is an appropriate balance within that mix?
- Who will implement the systems and solutions?
- Who will provide oversight? Decision making? Who will finance what part or all of the system?
- Are any new authorities needed with respect to enforcement, rules of operation, or other decisions?

Research into governance options will be informed by the technical and policy research described in this document. An initial step was taken with the development and hosting of a Governance Roundtable on June 20, 2011. Experts from other industries were invited to discuss the steps in analyzing what governance is needed with the introduction of new technologies and systems. Experts provided insight into existing models that could inform options for governance for the connected vehicle environment. Proceedings from this event are posted on the ITS Program website and form the basis for developing a strategy for next steps.

SECTION V. CONCLUSION

This document presents the options that form the basis for an approach to V2V/V2I communications security. The options were derived from an analysis of risks and existing industry best practices. The proposed approach is a combination of technical and policy options that results in a trusted and secure system for V2V/V2I communications and users, and one that adheres to specific conditions for privacy and scalability. The proposed approach reasonably meets these objectives and further considers the balance required in prioritizing safety and security requirements with costs, privacy, and institutional and governance requirements.

In summary, it is possible to combine hardware, software, and policy measures into an approach that can mitigate high impact attacks, making attacks highly improbable given cost and level of effort required for a meaningful attack, and thus assuring acceptance by users for trusting the system and for using applications for crash-avoidance safety. The known advantages of the approach are that it:

- Meets the objectives of providing trusted, anonymous messages using random identifiers that are changed every five minutes
- Is scalable to 250+ million users
- Supports crash avoidance safety applications
- Many attacks are only feasible with a significant amount of investment and expertise about the system
- Approach prevents/mitigates against harm to the system.

Some of the known limitations that will be addressed through further policy research to examine options include:

- There is no instantaneous identification of misbehaving actors. Inevitably, there is a delay in identification and delay in removing misbehaving actors from system.
- Splitting the certificate management entity may have cost implications.
- Currently, there are no clear financial models that offer a sustainable foundation for ongoing operational and maintenance costs.

APPENDIX A: DETAILS ON POLICY RESEARCH

A.1. Policy Research for Certificate Management Entities

The research into certificate management entity(ies) will result a review of best practices and lessons learned as a basis for tailoring an entity to meet V2V/V2I requirements:

- An evaluation of whether a Certificate Authority is the most appropriate type of entity to provide the communications security, certificate management responsibilities or whether there are other types of certificate management entities for consideration. This will include evaluation of relevant best practices and a comparison of entity types, discussing advantages and disadvantages of each.
- The challenges and risks to developing an operational/organizational model for a CM Entity under the existing Communications Security approach.
- Policy issues and trade-offs to consider when designing a CM Entity and a certificate management scheme (e.g. privacy, security, institutional issues, governance, and other key topics).
- An investigation into investigating varying institutional designs, ownership structures, decision-making processes and policies. This research will focus on the benefits, costs, and risks associated with wholly public, wholly private, quasi-public, or public-private partnership structures and evaluate how these varying types of organizations would be able to meet identified objectives of the communications security approach. It further includes a discussion of the Federal and/or government role in each of the organizational types.

The initial review will provide a basis for developing organizational structures and options from a physical operations perspective (i.e., centralized versus decentralized, or regional); from a cost perspective and ease of implementation (including impacts on the public sector and on industry), and from a cross-border perspective. Roles and responsibilities will be defined and will include (but not limited to) options for who is suited to:

- Manage user access and certificate issuance and to detect misbehavior and revoke certificates or distribute certificate revocation information;
- Store and retrieve data for misbehavior detection purposes and to backup databases ;
- Perform or provide security – physical and logical access;
- Establish and monitor system performance metrics and audit policies and procedures;
- Manage user privacy protection and provide enforcement; and
- Provide system administration and maintenance.

The models that result from the research will include proposed rules of operation/use, standard protocols, operational and decision making policies. These rules and policies are expected to include (but are not limited to) options for:

- Standard operating rules and protocols;
- Procedures for certificate issuance – how to register users; generate, sign, encrypt, and manage certificates; and manage revocation functions and information;
- Procedures for identifying, processing, and handling certification revocation and enforcement options for addressing and deterring misbehavior, including legal limitations of enforcement, and options for addressing poor system performance or failing devices;
- Decision-making processes and roles for making continuous improvements, changes, and updates to the rules of use, standard protocols, and operational and decision making policies;
- Backup plans and identification of disaster recovery levels required by functional objectives; and

- An analysis of how rules of operation/use are impacted when operating nationally, i.e., an entity operating across multiple states may face specific multijurisdictional and other policy issues.

Policy research will also provide options for rules of access to the certificate management system and proposed policies for all system participants: users, general public, system administrators, entity staff, deployment agencies, and the U.S. DOT. Rules of access are expected to include (but are not limited to) proposed options for:

- Processes for data administrators to securely access and manage data within the certificate management system and policies and processes for obtaining access for different system participants
- Policies that protect the system from unauthorized users; and
- Policies and processes for specific types of requests for access (e.g. law enforcement requests, requests from government agencies) and analysis of whether and how such requests might conflict with the privacy principles.

Ultimately, the policy research will result in an ability to construct full-scale organizational and operational models for real-world operating conditions. These models will include, but are not limited to, analysis of:

- Resources requirements including staffing by functional skill sets; facilities; equipment; and operating requirements.
- Costs including staffing, facilities, equipment (hardware, software, etc.), operations (daily, annual costs, etc.), maintenance (of equipment, etc.), administrative costs, other ongoing/recurring costs, and overall implementation costs.
- Implementation requirements which are expected to include an assessment of implementation steps and timelines (short and long-term milestones); an identification of task owners (roles/responsibilities) in implementation; resource requirements (level of resources needed and associated timelines/costs); authority/legal requirements (if necessary); and risks, challenges, and institutional barriers to implementation.

A.2. Policy Research for Communications Data Delivery System Options

Research that is planned or underway includes development of **options and business models for the communications data delivery system** for which communications security is required. Research includes an analysis of:

- The effectiveness and cost-effectiveness of various infrastructure and non-infrastructure configuration options for meeting requirements, particularly the requirement for frequency of security credential materials distribution;
- The capability of meeting key stakeholder requirements, particularly the security, reliability, and privacy required when distributing and managing credentials;
- The ability to meet performance metrics for coverage, reliability, redundancy, frequency of security credential updates, and ability to identify misbehavior, at a minimum;
- The acceptability of each option with respect to security; privacy principles; Federal, State, regional, local, and cross-border laws and regulations;
- The financial sustainability of each option, including the ability to leverage existing systems; the mechanisms that offer sustainable revenues in support of lifecycle costs (capital, operating, and maintenance); and the opportunity to attract investment partners or revenue sources;
- The challenges and opportunities associated with extending a communications delivery system primarily focused on communications for security credentials distribution and management to provide support for other applications including V2I safety, mobility and/or other uses of the connected vehicle environment; and
- The ability to implement and sustain the business model.

End Notes

[1] The research is administered through the U.S. Department of Transportation's (US DOT) ITS Joint Program Office (ITS JPO) and conducted in partnership with the Department's surface transportation modal administrations. The five year research agenda is described in the **ITS Strategic Research Plan**, **2010-2014** (http://www.its.dot.gov/strategic_plan2010 _2014/ index.htm).

[2] More detailed information on the transformative nature of these cooperative systems can be found in ITS Strategic Research Plan, 2010-2014 (http://www.its.dot.gov/strategic_plan2010_2014/index.htm and in the report, Frequency of Target Crashes for IntelliDrive Safety Systems at: http://www.nhtsa.gov/DOT/NHTSA/NVS/Crash%20Avoidance/Technical%20Publications/2010/811381.pdf.

[3] The Crash Avoidance Metrics Partnership (CAMP) provides an OEM-oriented research consortium under which various stakeholders can collaborate as desired on pre-competitive crash avoidance research projects of mutual interest. The VSC3 Consortium, consisting of Ford, General Motors, Honda, Hyundai/Kia, Mercedes, Nissan, Toyota, and Volkswagen/Audi, was formed under the CAMP agreement to conduct pre-competitive research on 5.9 GHz DSRC cooperative safety technologies and applications.

[4] The security teams are as follows: Carnegie Mellon University; the University of Illinois at Urbana-Champaign and Telcordia; GM India Science Labs; Security Innovation and eScrypt.

[5] From a technical perspective, this expansion will likely require further research on the integration of multiple communications platforms as well as the communications security of the data delivery from users and/or infrastructure nodes to back office systems. From a policy perspective, this expansion will require further research on privacy and governance.

[6] The research is administered through the U.S. Department of Transportation's (US DOT) ITS Joint Program Office (ITS JPO) and conducted in partnership with the Department's surface transportation modal administrations. The five year research agenda is described in the ITS Strategic Research Plan, 2010-2014 (http://www.its.dot.gov/strategic_plan2010_2014/index.htm).

[7] More detailed information on the transformative nature of these cooperative systems can be found in ITS Strategic Research Plan, 2010-2014 (http://www.its.dot.gov/strategic_plan2010_2014/index.htm and in the report, Frequency of Target Crashes for IntelliDrive Safety Systems at: http://www.nhtsa.gov/DOT/NHTSA/NVS/Crash%20Avoidance/Technical%20Publications/2010/811381.pdf.

[8] The Crash Avoidance Metrics Partnership (CAMP) provides an OEM-oriented research consortium under which various stakeholders can collaborate as desired on pre-competitive crash avoidance research projects of mutual interest. The VSC3 Consortium, consisting of Ford, General Motors, Honda, Hyundai/Kia, Mercedes, Nissan, Toyota, and Volkswagen/Audi, was formed under the CAMP agreement to conduct pre-competitive research on 5.9 GHz DSRC cooperative safety technologies and applications.

[9] The security teams are as follows: Carnegie Mellon University; the University of Illinois at Urbana-Champaign and Telcordia; GM India Science Labs; Security Innovation and eScrypt.

[10] In this document, a user is defined as all end users or users that send and receive messages through the System.

[11] Definitions located at: http://searchsecurity.techtarget.com/definition/PKI; http://en.wikipedia.org /wiki/Public_key_infrastructure; and in Cryptography Decrypted by H.X. Mel and Doris Baker, Addison-Wesley, December 2000.

[12] The required range is a 300 meter radius.

[13] The BSM is described by the SAE J2735 standard, part 1 which is likely to include a vehicle's temporary ID (instead of containing PII), time, position latitude, longitude, elevation, vehicle speed, transmission state, heading, steering wheel angles, acceleration, yaw rate, and break system status. It will also include data on a vehicle's path and current vehicle dynamics. Most data is measured by a vehicle's sensors. Time and location are derived from the vehicle's GPS signal. The final requirements will depend on the results of scalability testing at a later date. For more information on J2735, see: http://www.sae.org/standardsdev/dsrc/DSRCImplementationGuide.pdf.

[14] Bogus messages are defined as messages with an invalid signature or authentication tag.

[15] Risk = (impact to the system) x (likelihood of a successful attack).

[16] Application Programming Interface.

[17]Static identifiers include: certificate identifier; source address (at any level); identifier in application message payload; and vehicle characteristics, such as length, weight and vehicle type

[18] Dynamic identifiers include a vehicle's location, velocity, and acceleration.

[19] A physical level of hardware security will need to be determined for RSEs and ASDs.

[20] A Standard Controller and a Standard Controller with Security Features (controller has security features as part of the architecture, i.e. ARM's TrustZone offers virtual processors that cannot interfere with each other) were identical in terms of cost of an attack to the system and the time required for a successful attack to be mounted. The only difference was cost of implementation.

[21] More specifically, the controllers would share the CAN bus. A Can bus is a vehicle bus standard designed to allow microcontrollers and devices to communicate with each other within a vehicle without a host computer.

[22] Software would identify both as misbehavior because the system cannot determine the cause of a defect. The failure rate assumption used in the security research considerations was 1 failure in 1,000,000 hours of operation. Almost all technical defects would be able to be detected by the on-board diagnostics, and consequently a vehicle would be able to avoid sending out messages that would be judged to be misbehavior due to a technical defect. In combination with plausibility testing, the likelihood of a technical defect without the sender noticing it is extremely small. Precise numbers are not available.

[23] An ECU is a generic term for any embedded system that controls one or more of the electrical systems or subsystems in a motor vehicle. Many vehicles have over 70 ECUs.

[24] Bogus messages use an invalid signature, and spurious messages include a valid signature but incorrect payload data.

[25] Mobile users include any user device, such as a vehicle device (as described in this approach), pedestrian smartphones, etc.

[26] Static identifiers include: certificate; source address (at any level); identifier in application message payload; fields that may appear in the BSM (such as vehicle data).

[27] An attacker may also attempt to use the vehicle's Basic Safety Message (BSM) to track a vehicle. The BSM contains static identifiers – such as the vehicle's temporary identifier (in place of using PII) and a vehicle's physical characteristics. These properties assist to build a dynamic state map of the driving environment, which is used to predict the future movements of the vehicle, resulting in potential warnings to surrounding vehicles. More research will be conducted during the Safety Pilot to determine the trade-offs between breaking an attacker's ability to perform linking (by not including fields such as precise length, weight or vehicle type data), and the safety risks posed by breaking a vehicle's ability to building an accurate dynamic state map.

[28] The time between two consecutive times a vehicle communicates with a CA is called Time-To-Access (TTA).

INDEX

A

B

C

D

E

F

S

T

U

V

W